BITS of Belonging

Simanti Dasgupta

BITS of Belonging

Information Technology, Water, and
Neoliberal Governance in India

TEMPLE UNIVERSITY PRESS
Philadelphia • Rome • Tokyo

TEMPLE UNIVERSITY PRESS
Philadelphia, Pennsylvania 19122
www.temple.edu/tempress

Library of Congress Cataloging-in-Publication Data

Dasgupta, Simanti, 1971–
    BITS of belonging : information technology, water and neoliberal
governance in India / Simanti Dasgupta.
        pages cm
    Includes bibliographical references and index.
    ISBN 978-1-4399-1258-4 (hardback : alk. paper) —
ISBN 978-1-4399-1259-1 (paper : alk. paper) — ISBN 978-1-4399-
1260-7 (e-book)   1. Computer software industry—Social aspects—
India—Bangalore.   2. Computer software industry—Economic
aspects—India—Bangalore.   3. Information services industry—
Social aspects—India—Bangalore.   4. Information services
industry—Economic aspects—India—Bangalore.   5. Water-supply—
Economic aspects—India—Bangalore.   6. Water-supply—Social
aspects—India—Bangalore.   7. Social structure—India—Bangalore.
8. Neoliberalism—India—Bangalore.   9. Political participation—
India—Bangalore.   10. Bangalore (India)—Social conditions.
I. Title.
    HD9696.63.I43B364 2015
    338.4'7005095487—dc23
                                                        2015005951

♾ The paper used in this publication meets the requirements of the
American National Standard for Information Sciences—Permanence
of Paper for Printed Library Materials, ANSI Z39.48-1992

Printed in the United States of America

9 8 7 6 5 4 3 2 1

For Ma, who left too soon

Contents

Acknowledgments

Than and traveled with me for quite some time. It also brings closure, both professionally and personally. The moment has now arrived to thank all those who made this journey possible, productive, and pertinent. I would like to begin by expressing my deepest gratitude to Adriana Petryna, whose unwavering commitment as my thesis advisor, to put it simply, was astounding. Her critical eye helped me refine the arguments and value the ethnography not only as a methodological tool but also as a way to think about the ethics of fieldwork. My thanks go to Ann Laura Stoler for her wisdom, intellectual rigor, and relentless inspiration. By encouraging me to think at the intersection of anthropology and history, she instilled a new habit of inquiry in me that I have actively pursued since then. This research would not have been what it is if Gyan Prakash did not write *Another Reason*, the book that deeply inspired my ethnography with information technology professionals in contemporary India. I thank Gyan Prakash for agreeing to be on my dissertation committee and helping me think through the various issues of this work. The encouragement I received from Hugh Raffles and Richard Wilson through the years is invaluable. The camaraderie and support of friends I made during my time in graduate school is irreplaceable. I want to thank Erin Koch, Anne Galvin, Karolina S. Folis, Amanda Coleman, and Leo Coleman for being there. I am grateful to Supriya RoyChowdhury and Narender Pani for their deep insights and sustained comradeship. My sincere thanks to my editor, Aaron Javsicas, for his enthusiasm and support in bringing this book to fruition. I am deeply grateful to the (anonymous) reviewers for their insight and critical feedback, which has enhanced the quality of the work. I am deeply appreciative of the staff at Temple University Press and the production company whose work made this book a "thing."

The National Science Foundation, the American Institute of Indian Studies, and the New School for Social Research fellowships funded this work. I am grateful for all the institutional support without which this research would have been difficult to conduct. Bangalore became my home for a long time during this work, and several organizations and people I came to know made the stay meaningful and enjoyable. First, I am deeply indebted to Infosys Limited for allowing me to conduct fieldwork over short summer visits and then for an extended period of time. I am especially appreciative for the access I was granted to converse with employees at all levels of the company. I want to thank Ramesh Ramanathan and Swati Ramanathan of Janaagraha for letting me work with the GBWASP team to understand the water project in depth. I thank the GBWASP team for accepting me into the group. At Janaagraha, I developed a close friendship with Chitra Narayan (Chitra, if you are reading this, please contact me) and her family who welcomed me into their home; their love and support were irreplaceable. I am grateful to the support I received for this work from state agencies such as the BWSSB, KUIDFC, DMA, and the CMCs and TMCs where I showed up constantly with questions. Above all, my heartfelt gratitude to the residents of Manjunathanagara, who welcomed me into their homes and lives in ways that went well beyond our water conversations. Bina Chaudhuri, who left us, and Suparno Chaudhuri let me relive my childhood in their love and care through my stay in Bangalore.

By the time I was completing the manuscript, I had accepted my current position at the University of Dayton, where my colleagues welcomed me, especially Kristen Cheney and Shawn Cassiman, for which I am indeed grateful. I owe special thanks to Jamie Longazel, my departmental colleague and comrade, who has been an ardent inspiration while working alongside me on his own manuscript. I want to thank Ellen Fleischmann, Caroline Merithew, Haimanti Roy, Marybeth Carlson, and Jayne Robinson for their support and friendship as we all forge a life together at the university.

My parents did not live to see this book. To them, I owe my sincerest gratitude for providing me with what my father loved to call "the initial velocity." The initial velocity was significant in my life, not only to achieve my academic goals but also to carve a life outside the patriarchal norms of what it usually means to be a "woman." In retrospect, I applaud their courage to bring up a *child* rather than a daughter in a society entrenched in patriarchy. Tapas Ranjan Dasgupta, Anindita Dasgupta, Arijit Dasgupta, and Aparajita Dasgupta have stepped in to fill the void my parents left behind. Thank you for still making Calcutta feel like home. My childhood friend, Mahuya Pal, and her partner, Ambar Basu, moved to the United States while this work was in progress. Thank you Mahuya and Ambar for your sustenance in my darkest days. I am forever indebted to Prasanta Roy for encouraging me to lead the life of the mind. His humble counsel that as academics we are *sabhyatar thelawala* has protected me through the years.

The little girl without whom this research and this book would have been unattainable is Mrinmoyee/Putun. Her kindness, her indulgence, and her silent endurance assure me of eternal sunshine in my life. My closing thanks to Adam Williams, who came into my life at a time least expected. His presence has made my everyday meaningful in ways that I never thought were possible. Best of all, his ethics, his integrity, and his intellectual rigor have reinforced my faith in humanity and that all is not lost. Above all, it is my life with Adam and Mrinmoyee that I treasure the most and this book is a sign of that contentment.

Abbreviations

BATF	Bangalore Agenda Task Force
BWSSB	Bangalore Water Supply and Sewerage Board
CMC	city municipal council
CMM	Capability Maturity Model
DMA	Directorate of Municipal Administration
E&R	education and research
GBWASP	Greater Bangalore Water and Sanitation Project
IFC	International Finance Corporation
IT	information technology
KM	knowledge management
KUIDFC	Karnataka Urban Infrastructure Development and Finance Corporation
MoU	memorandum of understanding
O&M	operation and maintenance
PPP	public-private partnership
RWA	Resident Welfare Association
TMC	town municipal council
ULB	urban local body
WSP	Water and Sanitation Program
WTP	willingness to pay

BITS of Belonging

Introduction: Beyond IT

In 2002 during my first fieldwork trip to Bangalore, I met Nandan Nilekani, then the chief executive officer and one of the founders of Infosys Limited, the site of my fieldwork.[1] He argued that the best evidence of the emergence of India as a "powerhouse" is visible when software engineers migrate to the West. He presented a particular vignette: "When my boys go to the U.S. to work on a client site, they go through the immigration process with pride because even the immigration officer knows that he is not there to work at a gas station or a convenience store. He is there because the U.S. needs him to program software for them." The skewed gender and class nature of the vignette aside, what this points to is the emergence of a new nationalistic narrative.

Such narratives were also palpable in the IT company I worked with, which was headquartered in Princeton, New Jersey, with an offshore office in Hyderabad, in the southern Indian state of Andhra Pradesh. The majority of the software engineers and developers were Indian, mostly on H-1B work visas from India. When my Indian colleagues learned about my research topic, they encouraged me by pointing out that I was doing something "valuable" for our country. They collectively agreed that if there is "one thing" of phenomenal importance in India now, it is the rise of the Indian information technology (IT) industry on the global scene. They were also gratified that software engineers like them would be at the center of my work. One notion

---

1. Infosys, one of the leading IT companies in India, was established in 1981 and earned $8.25 billion in revenue in FY 2014.

that seemed to consistently underscore these preliminary conversations was that the IT industry flourished in a country otherwise relegated to underdevelopment, poverty, hunger, and corruption. Therefore, it is especially important that such success be celebrated and owned as a national pride.

It is rather interesting to note that the "pride" is primarily based on derivative work; that is, the majority of software development is designed to support the business processes of corporations in the West. The Indian IT industry has little to do with innovative work or program development. For instance, at the time of my fieldwork, Infosys had only one product, Finacle, developed for banking solutions. In February 2014 Infosys set up a subsidiary, EdgeVerve Systems, to concentrate its products, platforms, and solutions. The products and platforms currently account for only 5.2 percent of the $8.2 billion of the revenue of the company.[2] When I wanted Nandan Nilekani to explain the dearth of investment in products, he mentioned that the company's expertise lies in software development, and that is what Infosys is known to do well. In this context, Xiang Biao's *Global Body Shopping* examines a possibly unique Indian practice called "body shopping" in the IT trade between India and other countries, especially in the West. Body shopping is a practice where "an Indian-run consultancy (body shop) anywhere in the world recruits IT workers, in most cases from India, to be placed out as project-based labor with different clients" (Biao 2007: 4).[3] While Infosys is no longer a primarily body-shopping company, this practice lends perspective to the Indian IT industry in the global context. As Biao argues, it is noticeable that software development is also premised on low-priced, English-speaking labor available in India. These kinds of global practices are inherently produced through the realignment of historic unequal global relations. Second, the IT process rather than embedding Indian IT professionals in the global circuit in effect makes them more vulnerable, hence disembedding them.[4]

---

2. "Infosys Hives Off Products Business into Separate Unit," *Times of India*, May 12, 2014, available at http://timesofindia.indiatimes.com/tech/tech-news/Infosys-hives-off-products-business-into-separate-unit/articleshow/34989367.cms (accessed September 26, 2014).

3. For a detailed discussion about the rise of the Indian IT industry in the national and international perspective, see Heeks (1996), Kumar (2001), Lakha (1990, 1994), Lateef (1996), Parthasarathy (2004), Saxenian (2000), and Upadhya (2004).

4. A conversation about the importance of innovation to sustain the Indian IT industry in the global market is slowly emerging, led by the National Association for Software and Services Companies (NASSCOM), the trade association for the industry. This is partly fueled by the economic downturn and partly by other countries rising as software development destinations. For details, see "Innovation—A Call for Action for the IT/BPM Industry," available at http://www.nasscom.in/sites/default/files/uploads/mailers/2013/CEO_Forum_Decoding_Innovation_Post_event_24052013/docs/Decoding_Innovation_17May2013_Synopsis.pdf (accessed October 2014); and "Innovation Transforming the Growth Landscape: Capitalizing IP in Indian IT/BPM," available at http://www.kpmg.com/IN/en/IssuesAndInsights/ArticlesPublications/Documents/Innovation-Transforming-the-growth-landscape.pdf (accessed October 2014).

Notwithstanding the shallow foundation on which the pride of Indian IT is based, the narratives are nonetheless vital ethnographic data. It is my argument that the IT pride is not limited to economics alone. It further, and perhaps more importantly, incorporates a strong ethical dimension. My colleagues at work narrated the success of IT as an ethical watershed for the nation-state. Adjectives such as "honest," "good," "not corrupt," "reliable," "educated," "good background," and so on, were often evoked to describe IT professionals, especially high-profile entrepreneurs.[5] For a country, especially the state, generally known to engage in corrupt practices, where global business would not venture easily, things changed with IT; they argued, "We have become more trustworthy as a country." In this view, the IT industry has carved out an ethical space that is distinct from and in disdain of the "corrupt" state. Two words that I encountered ceaselessly, which, as software professionals often reminded me, were at the core of global corporate governance and the market ideals, were "accountability" and "transparency." The market rhetoric therefore goes beyond sheer economics; it is also a normative tool that ensures uncompromising honesty in economic transactions.

What does an ethnography of neoliberalism look like? This book is an effort in that direction. It is about the narratives, discursive practices, and contestations through which a neoliberal critique of public governance emerges, thickens, and is also subverted in contemporary India. The critique plausibly emanated with the success of the IT industry, but, as I demonstrate, it is more importantly a politically and ethically negotiated narrative that coalesced over time to become an identifying account mainly of the urban middle class. I emphasize the class characteristic of the narrative to show how, on one hand, neoliberalism privileges the middle class, and, on the other hand, also provokes a different kind of politics of rights among the urban poor. I argue for a constitutive relationship between the ideology of the IT industry and the emerging neoliberal ideology of the polity without privileging either. Hence, my attention here is on simultaneity rather than on causality. Simultaneity compels us to look beyond boundaries, of IT in this case, to explore sites that are now being folded into the neoliberal schema. This book is therefore also about going *beyond* IT. The "beyond" in this instance is a water project, known as the Greater Bangalore Water and Sewerage Project (GBWASP; henceforth, Water Project), that would not only privatize water supply but also required a beneficiary capital contribution from citizens, often referred to as "customers." Water offers us the kind of "beyond" that unsettles the otherwise euphoric national narrative of IT. This "beyond" corroborates that the narrative is inherently embedded within and asymmetrically privileges the middle and upper-middle classes.

---

5. Smitha Radhakrishnan examines "background" in detail as a key word in the IT industry; she argues that the term oscillates between "claim to universality and relativity" (2011: 8).

In this newfound basis of class politics, belonging in the nation-state is once again fractured into bits, simultaneously familiar and unfamiliar. Bits in its metaphorical extension—BITS, as in the digital vernacular—is a binary, it takes the value of either 1 or 0. Binary digits store information and are the basic unit of communication in computing. A logical value of 1 is considered a true statement, and 0 is a false statement. The market paradigm post-1991 is parallel to the 1 BIT as the logically true statement. It leaves the other existing option, the socialist-redistributive model, 0, as a false choice. In a modernist model such as this, the choice is obvious, legible, and privileged. Thus one gradually ceases to belong to the nation-state as a citizen based solely on birthright. Belonging is based on an active participation in the market and also in accepting the market as the only viable option for the polity to prosper. In another sense, it is the acceptance of the logical value 1 as the true statement of the BIT and discarding 0 as obsolete.

I gathered success stories through ethnographic work with IT professionals in Bangalore, the Silicon Valley of India, and collectively call them the "IT narrative." The establishment of an Indian industry in the knowledge economy otherwise dominated by the West had an effect beyond fiscal reports. It has instilled a sense of confidence among the IT professionals that they can now offer better leadership to the country based on their experience of competing and succeeding on a global scale. My analyses of the success narrative of the IT industry generated the following set of values and principles espoused by entrepreneurs and software engineers. First, the IT industry in India succeeded not merely for the software knowledge and English-language skills available in the country but also because the IT companies meticulously adhered to the ethical corporate practices stipulated by global capital. Second, the success of IT offers a blueprint for governance, which can and *should* inform both private and public sectors. Third, the success narrative has lifted India from its enduring status as a poor developing country, and this opportunity ought to be utilized for a nationwide governance and institutional reform, especially for the state. Fourth, the reforms can be effective by incorporating citizens, who believe in the power of the market as a transformative tool, within the public administrative apparatus.

Together, I term the above set of values the *ethico-political* ideology of the IT industry to emphasize its ethical nature, but also to indicate how the emergent set of ethics has established a new politics of belonging to the nation-state. While this book deciphers the cultural drivers and historical contingencies that underscore and manufacture the IT narrative, more importantly it problematizes the cohesiveness and legitimacy of the narrative itself. The apparent tidiness of the narrative intrigued me for two reasons: first, its seemingly unproblematic passage from the economic/corporate/capitalist to the social/public governance realm. I ask what enables these successful corporate entrepreneurs to offer guidelines for renewing the nation, which has two distinct domains—one relies on the market, the other on a fictive bond connected by birth and land. The question is not only about the passage of this narrative but also what makes IT possible at this particular historical juncture. Second, and I think more importantly, when we realize that the IT success narrative is one primarily of and by the middle class, it starts to fray at the edges. How does one account for class differences in a narrative that favors the market ideology? How does one understand the disenfranchisement of the poorer sections of the citizenry who are unable to participate in the market paradigm of change? What kind of class politics does this narrative provoke and suppress? In other words, who becomes the privileged citizen of the nation is now a question of whether one is "inside" or "outside" IT.

The inside and outside, as I show in this work, does not necessarily mean that one is employed in the IT industry. Relatively, it is an indication of an ideological divide of those who share and participate in the market ideals and those who do not, or rather cannot. Using ideology to depict the set of values is at once a simple and a difficult task. Linking this ethico-political narrative to the middle-class and upper-middle-class IT entrepreneurs and professionals is to suggest that ideology is "a system of beliefs characteristic of a particular class or group" (Williams 1977: 55).

In the Marxian sense, ideology arises out of the material processes that organize economic relations, but it is nonetheless a "false consciousness." Engels explained, "Ideology is a process accomplished by the so-called thinker, consciously indeed but with a false consciousness. The real motives impelling him remain unknown to him, otherwise it would not be an ideological process at all" (quoted in Williams 1977: 65). Raymond Williams has argued that the separation of the consciousness as a "reflective" or a "second" stage poses a difficult problem for conceptualizing ideology since consciousness itself, he argues, is formed through material social processes (Williams 1977). While Marx underscored the importance of material and social history that enhanced the concept of ideology, moving it beyond scientific empiricism proposed by Destutt de Tracy, his negation of "what men say" and "men as narrated" or from language itself, as Williams contended, created an enduring conceptual uncertainty (Williams 1977). In emphasizing the IT narrative, I draw attention to ideology not only as emerging from the material social relations of software development, a process that spans the globe, but also what the IT entrepreneurs say and how they are being narrated, for instance, by the media, as the harbingers of the neoliberal market. Finally, how does language come to delineate the new set of values and principles or, in other words, the IT ideology?

In the context of a class analysis, a couple of characteristics are unique to the IT industry in India: First, its emergence independent of the older business model that was linked to family businesses or established corporations; second, the educated middle class, who were the main founders of IT companies, drew on their social capital to establish a different corporate norm (Upadhya 2004). Both these characteristics were reflected in the narrative alongside class privilege. Further, Leela Fernandes has argued that there is a need to move beyond the obvious contribution scholars have made either "estimating the size of the middle class" or "analyz[ing] the middle class through the lens of consumption" (2006: xvii). Fernandes instead explores the political routes through which an entrenched connection among the middle class, consumption practices, and liberalization evolve while simultaneously arguing that the middle class is not a monolith.

Fernandes's invitation is useful because it leads us to look at processes through which the middle class garners, retains, and normalizes emergent ideologies and practices. Therefore, what troubled me about the connection that I heard being formulated among IT professionals, IT's middle-class proponents, and the nation-state was twofold. First, the narrative leap: If IT benefits the educated urban middle class, somehow it will be beneficial for India as a whole. Second, related to this, the high acoustics of the IT narrative was drowning out issues of inequality that were being introduced by the same 1990s liberalization policies that enabled the ascent of IT. Where do people from the lower social economic strata find their place in this changing economic terrain? Or, seen from a slightly different angle, if IT contributes 9.5 percent to the GDP,[6] how does it become relevant for the entire nation relative to others sector with higher contributions, for instance, the manufacturing sector (15.2 percent in 2012–2013)?[7] Moreover, even though the IT entrepreneurs I knew at Infosys upheld the class mobility offered by the industry, the education and soft skills one needed to secure a job in IT were middle-class privileges. Consequently, as I started this work in 2002, it soon became apparent that to tell the "success" story of Indian IT, one had to move "outside" the industry.

The resulting matter that arises here is relationship among IT, the state, and neoliberalism. Would IT succeed without neoliberal state policies? Perhaps. Does the visibility of IT exclusively rest on a "failing" state? This question demands a more nuanced exploration. The contradistinction erected between the ethics of the IT narrative and the "corruption" of the state cannot be simply relegated to the chronological "failure" of the state and the "success" of IT; neither can we see the latter merely as a direct outcome of the former. Rather, the field of possibility within which the IT narrative rose and gained traction has to do with the confluence of global and national processes that began propagating the market over socialist-redistributive state agendas.

6. NASSCOM, "Impact on India's Growth," available at http://www.nasscom.in/impact -indias-growth (accessed April 6, 2015).

7. Available at http://articles.economictimes.indiatimes.com/2013-05-12/news/39203854_1 _assocham-survey-100-million-jobs-manufacturing-sector (accessed May 10, 2013).

Undoubtedly, the Indian state, especially under the leadership of the Congress Party, has had a protracted and legendary scale of corruption, but to argue that IT's visibility solely depended on this is to miss the larger forces of neoliberal change that are sweeping the globe. The 2014 election is a case in point. The Congress Party lost the elections by a significant margin to the Bharatiya Janata Party (BJP); Nandan Nilekani, who ran for office on the Congress ticket, lost. There are local- and state-level distinctions of electoral politics in India, but it is also important to attend to neoliberal forces that increasingly undermine the legitimacy of the state as suitable for public governance. Therefore, this book does not argue for causality between the "success" of IT and the "failure" of the state. Instead it is an invitation to examine the complexity of neoliberalism, especially in India starting in the 1990s.

Neoliberalism as a concept is neither succinct nor analytically constant. There is also the accompanying intellectual dread that while neoliberalism could be a useful concept to capture the field of possibility, it may also be diluted as a mere placeholder in contemporary anthropological research.[8] In a 2014 *Annual Review of Anthropology* article, Tejaswini Ganti identifies two ways in which anthropologists have primarily used neoliberalism: "one is concerned with policies and politics and the other with ideologies and values" (Ganti 2014: 94). I find David Harvey's formulation of neoliberalism especially helpful for my analysis here because it underscores the coupling of values and ideals within the public institutional framework.

Harvey writes:

> Neoliberalism is in the first instance a theory of political economic practices that proposes that human well-being can best be advanced by liberating individual entrepreneurial *freedoms* and skills within an institutional framework characterized by strong private property rights, *free market* and free trade. The role of the state is to create and preserve an institutional framework appropriate to such practices. (2007: 2)[9]

As we can see, freedom here does not quite relate to a democratic mandate and citizens' rights. Instead it is tied to the unfettered possibility of

---

8. In January 2013 the University of Manchester Group of Debates in Anthropological Theory discussed a motion titled "The Concept of Neoliberalism Has Become an Obstacle to the Anthropological Understanding of the Twenty-First Century." Interestingly the final vote went for support of the concept (Ganti 2014). Available at http://www.talkinganthropology.com/2013/01/18/ta45-gdat1-neoliberalism// (accessed October 6, 2014).

9. Emphasis mine.

securing private property, where the state *enables* a market structure that privileges the already privileged. "Freedom," as Harri Englund notes, "is an essentially contested concept. Philosophical debates aside, the exportation of 'freedom' across historical periods and geopolitical areas have given the term new meanings, which are often ill-understood by those who have a specific agenda to pursue" (2006: 2). This idea of freedom, in addition, creates a new subjectivity that dissociates from the collective to settle on the individual as its point of reference. Nikolas Rose has written that "freedom is the name we give today to a kind of power one brings to bear upon oneself, and bringing power to bear upon others . . . and freedom is particularly problematic when we demand to be governed in its name" (1999: 96). This emergence has deep implications for a democratic regime since it signals, contrary to what some scholars have suggested, not the end of politics but a politics that is admissible only when filtered through a preset script of resistance and response.

Ganti makes another valuable distinction, that is, between "late capitalism" and "neoliberalism," emphasizing two points of divergences. First, late capitalism is a chronological marker, whereas neoliberalism is an "ideological and philosophical movement . . . that emerges at a particular historical moment" facilitated by network of intellectuals (Friedrich August von Hayek, Walter Lippman, Milton Friedman, and others) and institutions (Mont Pelerin Society) (Ganti 2014: 91). Second, neoliberalism is characterized by a call for a prescriptive arrangement between the state and its citizens to reorganize capital and property (Ganti 2014). Neoliberal futures are hence aspirational and call for a discontinuity with past forms of welfare governance. The discontinuity is a normative stipulation for the market to masquerade as an impeccable logic of well-being. Above all, I take seriously Ganti's (2014) suggestion that neoliberalism as an analytical concept helps us ground ethnographic work to think through processes of globalization that are otherwise diffuse and obscure.

IT professionals demonstrated a high regard for the market rather than the state as an underwriter for their well-being or, as they frequently expressed it, "the good life." They often presented private property they had acquired, which was "unthinkable" for their parents, such as owning a flat by the time one was thirty, cars, high-end gadgets, shopping trips to the mall, dining at eclectic restaurants, international vacations, and so on, as markers of their well-being. Still, I often noticed that the narratives of personal success and possessions often elevated from the personal to the national level, somewhat surreptitiously and quite naturally. The intertwining of individual biographies and the national trajectory indicated a deeper belief in the potential of the market to alleviate India from the stigma of being a poor country. This, as Paul Treanor has suggested, reflects the "belief in the moral necessity of market forces in the economy, [and] is probably the first defining feature of market liberalism."[10] In this sense, well-being as an experience was not necessarily confined to the IT workers. It was seen as a possibility for the nation as a whole, realizable only through the state being invested in the mechanism of the market.

On the other hand, it was quite remarkable to see that in nearly all the conversations about "accountability" and "transparency," seldom did anyone refer to its familiar humanistic merit. They were regularly cast in terms of corporate values and practices, which were also apparently new to Indians. I found it challenging to talk about these values in the familiar humanist domain in my discussions. While most professionals I interviewed acknowledged that these terms have long underscored collective life, they were reluctant to see these values now outside the corporation. Part of the reason, some explained, is that with a balance sheet lurking at the end of every financial quarter, one can see the tangible results—profit and loss—and hence the moral urgency to be accountable and transparent. Others were skeptical that such values could even work outside the supervision mechanism of the market, such as the "dashboard," which I discuss later, that companies like Infosys use to track project timelines and employees' productivity. In other words, the corporate governance paradigm was enumerative and protective. As Treanor reminds us, "Neoliberalism is not simply an economic structure, it is a philosophy . . . neoliberals tend to see the world in terms of market metaphors."[11]

One manager at Infosys explained the global relevance of "accountability" and "transparency" to me: "These could be daunting for people who are used to corruption, but this is what we have embraced as our core values. This comes to us naturally," he continued. "IT has done an excellent job in showing that these values work in the global economy; otherwise why would big companies in the U.S. come to us?" Neoliberal changes have recast the developing world since the 1990s, and India is not exceptional in this regard. Nonetheless, the IT industry "naturally" espousing these values is a rather problematic statement in that a specific class with wealth and financial power is claiming a set of values as proprietary. Julia Elyachar begins her discussion of neoliberalism by stating, "Anthropologists have a good reason to hate neoliberalism" (2012: 76). However, rather than blaming neoliberalism for all that has gone astray in the contemporary world, she invites us to attend to the "political debates and conceptual conflicts" that underscore the present. Elyachar demonstrates the departures between Hayek's formulation of neoliberalism and the ethnographic work on the same issue related to public infrastructure

---

10. Available at http://web.inter.nl.net/users/Paul.Treanor/neoliberalism.html (accessed March 27, 2013).

11. Paul Treanor, "Neoliberalism: Origins, Theory, Definition," available at http://www.web.inter.nl.net/users/Paul.Treanor/neoliberalism.html, 9 (accessed April 6, 2015).

in Cairo. She links the two through the concept of "tacit knowledge," something shared by anthropology and neoliberalism. Contrary to Hayek's argument that tacit knowledge can only be harnessed in the free market and stems from the public sector, Elyachar discovered the opposite in Cairo. Similar to Elyachar's demonstration of the *fahlawah* (street smarts or trickery) as important to the success of the public sector, state representatives working in the Bangalore slums often had to draw on the existing network of favors and vocabulary within the slum to convey the value of the project. Taken together the above theorizations signify the complexity of neoliberalism rather than it being a simple account of dispossession. Therefore, I document the struggles that are inherent in the social margins and the collective knowledge that is evoked, based on past experiences, to confront the powerful.

Alongside "success," narratives of the "outside" were another related ethnographic data I collected during my fieldwork. It was an expression that IT workers habitually used to demarcate their lives and circumstances from those who continue to be external spectators of IT. The inside of IT is broader than the industry itself; it included the emerging middle classes who are participating in the neoliberal market in different professional capacities, such as in biotech, banking, real estate, and so on. IT in this sense stands as a metaphor for the arrival of the middle class. The outside comprises those who are unqualified to participate in or who are even undeserving of the "good life," mainly because of their economic dispossession, such as the "urban poor" in my research. The outside offers us an analytic opportunity to interrogate the picture of change embraced by IT. Stated differently, rather than looking from outside IT, the outside is the window to look *into* IT. The IT entrepreneurs at Infosys assertively told me that their commitment to social change is not limited to the industry and is not even about the industry any longer; according to them, by the late 1990s the industry had established an "equitable" relationship with the state. The state, they stressed, was aware that supporting the IT industry was critical for India to gain global recognition. Therefore, Narayan Murthy (cofounder and chairman-emeritus of Infosys Limited), Nandan Nilekani, and others insisted that I definitely look "outside" IT, which is critical to their agenda of instituting market reforms in the state apparatus. In this sense, the outside is in fact not incidental or merely adjacent to the inside—that is, IT—but is at its core. Yet, what came across in my conversations at Infosys is that they perceived the outside as a homogeneous domain that is malleable and easily amenable to market transformations. Such a portrayal raised several questions. How does one account for the dichotomy between IT and the outside? And why is the outside necessarily perceived as "homogeneous"? What is at stake in this homogeneity when we think of a neoliberal schema?

The "outside" nonetheless posed an ethnographic challenge: What could be an ideal research site? As I was seeking a possible site, Infosys invited me to attend a citizens' meeting to discuss issues that are affecting city governance. They thought I would be interested because Nandan Nilekani was one of the key speakers. I initially interpreted Nilekani's involvement as a well-known denizen of the city. However, when I heard him speak that day, it was obvious that he was promoting the market ideals that would streamline governance and improve citizen engagement. He stressed the value of "accountability" and "transparency" while talking about a water project in the city. Both these values that have guided his experience with corporate governance, he argued, would be useful in organizing public governance, as well. He urged the audience to rethink their relationship with the state, particularly in relation to basic amenities such as water. Following this he mentioned an impending water project known as Greater Bangalore Water and Sewerage Project (Water Project) and the role of Janaagraha, a nongovernmental organization (NGO).

Janaagraha was organizing the citizens' participation component of the Water Project that involved monitoring and holding the state accountable for any delay and/or corruption. The Water and Sanitation Program of the World Bank was funding this component. In the brief window I found during lunch that day, I introduced myself to Ramesh Ramanathan, the cofounder of Janaagraha, along with his wife, Swati Ramanathan. In order to gain a sense of the "unique" work his NGO is involved in, Ramanathan invited me to attend the "Monday Morning Meetings" at his organization, where they present updates on the various projects. He assured me that if I was interested in measuring change, his organization would be a good place since "we are a citizens' platform and work on re-engaging citizens with the state after a gap of five decades."[12] Nilekani and Ramanathan had worked together as part of the Bangalore Agenda Task Force (BATF). The BATF was a public-private initiative founded by the then chief minister of Karnataka, S. M. Krishna, to reform urban governance in Bangalore. Therefore, like Nilekani, Ramanathan trusted the market as the only tool that can bring about substantive changes both in the way the state functions and how citizens can involve themselves in governance as "stakeholders" of the nation. The Water Project, he stressed, would be a useful area to study the renewed engagement between the state and the citizens. At the end he said, "We do not need to reinvent the wheel if the market has succeeded in every other area." Attending the citizens' meeting that day led me to my "outside" infrastructural site.

---

12. At that time, the Ramanathans were reluctant to label Janaagraha as an NGO. They preferred to call it a "citizen's platform" to indicate its urban dimension and also to separate it from the developmental paradigm, which is often seen as the domain of NGO work. I have discussed this active delinking that the urban middle class has embraced elsewhere (Dasgupta 2012).

The study of infrastructure is not new but is developing as a more focused area of inquiry.[13] In 2012 *Cultural Anthropology* curated a volume, which dwelt on how scholars have started theorizing the subject of infrastructure through ethnographic work.[14] Theoretically this ranges from thinking of infrastructure as a "social-material assemblage" (Nikhil Anand) to how the ordinariness of infrastructure is "appropriated and recombined" by users (Jonathan Bach) and then how we can think of it through the mechanism of gift and reciprocity (Daniel Mains). Elyachar, on the other hand, is reluctant to theorize infrastructure yet because, she says, "At certain moments fuzzy concepts are more helpful to capture the number of interlinked issues."[15] My own theorization of water supply as a question of infrastructure relates to the above positions, but I am also interested in water being the paradigmatic case for the provision of basic amenities. Further, through this work, I underscore how the vocabulary of the market, such as transparency, rather ironically coincides with the nature of water.[16]

The Water Project was designed by the United States Agency for International Development (USAID) and primarily involved the privatization of water funded jointly by the beneficiary citizen contribution, market bonds, mega city loan schemes, and the government of Karnataka. The continuity of the market narrative between IT and the Water Project was remarkably seamless and thorough as though they were identical areas of human life. Throughout my fieldwork I found it quite arduous to engage people both at Infosys and Janaagraha to attend to the specious similarities being drawn between IT, a capitalist enterprise, and water, a basic and sacred amenity. Though some partially acknowledged the difference, their overarching idea was that by creating a water market one can end the water woes that plague the city, and water scarcity was definitely a major infrastructure issue in burgeoning Bangalore. Such awkward and unexpected interrelationships have led to the rise of tensions, especially around basic amenities, that reflect deeper schisms in a social order that was already premised on caste and class hierarchies. The depth of neoliberalism, as Harvey writes, lies in the belief that "if markets do not exist (in areas such as land, water, education, health care, social security or environmental pollution), then they must be created, by state action if necessary" (2007: 2). The Water Project, in this case, was a

---

13. Susan Leigh Star's work was perhaps the first to dwell on this subject directly. See Star (1999).

14. "Infrastructure: Commentary from Nikhil Anand, Johnathan Bach, Julia Elyachar, and Daniel Mains," Curated Collections, *Cultural Anthropology Online*, November 26, 2012, available at http://production.culanth.org/curated_collections/11-infrastructure/discussions/6-infrastructure-commentary-from-nikhil-anand-johnathan-bach-julia-elyachar-and-daniel-mains (accessed January 18, 2013).

15. Ibid.

16. See Dasgupta (2010).

state-initiated infrastructure project, but it was based on the public-private partnership model. Studying water, which is considered sacred in the Hindu imaginary, in its reincarnation as commodity opened up a new dimension to explore how the market shifts the meaning of citizenship and belonging to the nation-state. The new notion of belonging was gradually linking citizenship to consumption, where one can be acknowledged as a citizen only as a consumer of services and products.

Treanor makes a compelling argument in this context, that neoliberalism is "the desire to intensify and expand the market, by increasing the number, frequency, repeatability, and formalization of transactions."[17] To this formulation I would like to add that the desire for a new normativity based on the market was ethnographically deep. As I started charting the Water Project, besides the middle class I came to know slum residents, or "urban poor" as they were officially categorized, who would be included in the new initiative. The urban poor were now being asked to pay toward the project cost, which was adjusted according to the square feet of their dwelling and also for the water they would eventually consume. The urban poor characterized these payments as "unacceptable demands" by the state since historically they received free water as a public service to those below the poverty line. The resistance to the market ideals started mounting as I began my work in the slums.

Over the past decades, the involvement of NGOS and community-based organizations (CBOs) in development projects has become an established norm, which is often labeled as "participation" of the beneficiaries in making decisions about projects' designs and their implementation. International donor agencies, such as the World Bank, have been influential in setting this new paradigm in motion. Part of this initiative was meant to counter the critique of the top-down model of the projects. However, on the other side, with the spread of neoliberal values, the states in the developing world increasingly came to be portrayed as "corrupt" institutions with a severe lack of "transparency" and "accountability." This kind of dismal portrayal of public governance was disseminated by the agencies like the World Bank and was eagerly embraced by the rising middle class, particularly middle-class NGOs, who were looking for an ideological sponsor to confine the role of the state as a guarantor of the market economy. Scholars have noted the record rise of NGOs and CBOs recently in countries like India. Baviskar writes, "Among the social groups and associations that are considered to make up the civil society, NGOs have become especially prominent in the last two decades" (2001: 1). Thus, we need to ask whether we can substitute civil society in the Gramscian sense with NGOs' and CBOs' contributions. Though these organizations stand apart from and yet are able to impact the state, can they represent class

---

17. Paul Treanor, "Neoliberalism: Origins, Theory, Definition," available at http://www.web .inter.nl.net/users/Paul.Treanor/neoliberalism.html, 9 (accessed April 6, 2015).

interests equitably? Or do these emergences indicate a new form of class politics that is otherwise disguised in the name of the collective good?

Sharma and Bhide (2005), examining the World Bank–funded Slum Sanitation Program (SSP) in Mumbai, have argued that the participation of the slum dwellers was minimal despite the involvement of well-established NGOs, such as the Society for the Promotion of Area Resource Centers (SPARC). I encountered a similar situation while working in the slums of Bangalore, where one resident commented, to the ironic amusement of others, "Yes, we will participate, because we are good at it." Therefore, rather than rushing to evaluate "empowerment" achieved through participation, which we often see in development agency reports, it is important to examine the politics of "participation" itself.

Another recent issue that has surfaced in the development sphere is the increasing donor participation of countries like India, China, and Brazil, which were previously recipients of aid. As Emma Mawdsley has argued, the fracturing of what was previously a domain of the Western powers has led to a complex situation. She contends that this calls for a recalibration of "international relations and political economy of development" along with a departure from the traditional analytical binaries such as "North-South," "East-West," "developed-developing," "First-Second-Thirds Worlds" (Mawdsley 2012: 3). Elsewhere I have argued that Janaagraha's participation in the Water Project was intended to delink participation from the sphere of development and claim it for the market (Dasgupta 2012). As I discuss later, Janaagraha was even reluctant to include the urban poor as a separate module in the participation structure. Their primary argument was "citizens" is already inclusive and does not necessitate a separation by class. They finally agreed to do so only when the Water and Sanitation Program (WSP) of the World Bank made it mandatory for them to receive the funds. Such reluctance to involve the poor based on class location or even labeling the Water Project as a "development project" within a donor-recipient country like India, signals the contemporary complexity of the field that Mawdsley discusses in her work.

Based on the above theoretical framework, some questions that arose in my discussion with the slum residents were the following: What does it mean to pay to connect to the main water pipe? Isn't it part of the government's responsibility? Why do we have to pay for water now? What has changed? Is the government no longer going to take care of the poor? Finally, where is the water going to come from, since the river (meaning Cauvery) is so far away? These questions unequivocally displayed the core problems of market reforms. Together—the slum dwellers and I—were led to a space where the otherwise "impeccable" logic of the market endorsed by IT was fraying at the edges. Answers were difficult to come by, but the questions mattered. Even when some state actors were confronted with these questions, they were befuddled. Contrary to the expectation of the IT narrative, the state was not receding or reincarnating itself to make way for the market. Rather, what was emerging was a state that was muddling through the new set of practices that had descended on it.

My meeting with local councilors, who often used water to garner votes in the slums, revealed that they were grappling with the new ideology and government processes that they themselves barely understood. This was further demonstrated in my work with the Bangalore Water Supply and Sewerage Board (Water Board),[18] a para-statal agency bestowed with the task of supervising the implementation of the project. In several meetings I attended at the Water Board, engineers who had years of experience working in the slums of Bangalore were hesitant about charging the urban poor for water. The urban poor, on the other hand, were deeply aware of this kind of predicament faced by administrators and elected councilors. Nikhil Anand, in his work on water supply in Mumbai settlements, argues that strategies to secure water invite us to "move beyond binary theorizations of haves and have-nots that are commonplace in writings about cities, especially those in the Global South" (2011: 543). To think of reassigning water as a commodity under privatization initiatives therefore is not merely about an added financial burden for the urban poor. Such changes further indicate the recession of final frontiers, water in this case, that long enabled the urban poor to engage, even minimally, in the power structure. The urban poor resistance to water privatization is not archaic; it is, as Raymond Williams helps us reflect, "residual." The residual is not "archaic" because it not "wholly recognized as an element of the past" (Williams 1977: 122). In bringing the urban poor within the ambit of my analysis, I want to suggest that despite their otherwise marginal locations, their resistance is constitutive of the neoliberal discourse.

As I completed this manuscript, the contemporaneity that framed my ethnography somewhat changed. However, the changes, rather than negating, have come to reinforce the neoliberal system that I set out to study. There are four issues that I feel compelled to address to resituate this work. First, the current global economic downturn and its effect on the IT industry in India; second, the increasing role of IT entrepreneurs like Nilekani in public governance and seeking public office and losing in the 2014 election; third, the increasing middle-class "capture of the state" that is congealing as a wider urban middle-class ethos rather than being confined to organizations such as Janaagraha; and, fourth, the 2014 election itself that catapulted the Hindu fundamentalist party, BJP, to power at the center by a 53 percent

---

18. I still selectively use BWSSB while referring to formal acts and policies.

majority ending the coalition government phase, the norm for some decades now. Apparently these events seem disconnected and somewhat contradictory because Nilekani's defeat may seem to be the demise of the IT ideology that endorsed the market for public governance. However, when we consider that Narendra Modi, the current prime minister of India and the leader of the BJP, won on a platform of the "Gujarat Model of Development," which favors the neoliberal agenda and is protective of what Modi himself has termed the "neo-middle class," a more complex picture of the market-based ideology emerges. The ideology now signals a wider emergence of a new middle-class "capture of the state" to endorse marketplace consumerism. Further, as I show in Chapter 3, while IT professionals may not endorse aggressive Hinduism, sympathies for a lighter version of Hindu nationalism, or "Hindutva Lite,"[19] was quite common, just like it is among middle-class India now.

The recent global economic downturn has undoubtedly affected the IT industry in India since most of the clients of the industry are located in the United States and Western Europe. Based on these findings, *The Hindu* reported, "For the first time in 47 quarters, Infosys, hitherto the bellwether of the sector, missed its revenue growth guidance. Its guidance for 2012–13 is just 8–10 per cent growth in U.S. dollar terms compared with the industry association National Association of Software and Service Companies' (NASSCOM's) guidance of 11–14 per cent."[20]

Though Infosys reported increased revenue in the April–June quarter in 2014, according to the *International Business Times*, the future projects of the company are not as promising compared with the past since the company reduced its earnings per share guidance for 2012 to 2.97 from its $3.03 projection last year.[21] The *Wall Street Journal* further reported that countries such as Indonesia, the Philippines, and Mexico pose a serious threat to the outsourcing industry in India.[22]

Despite the economic slump, IT entrepreneurs continue to significantly carve out their space in public governance. As they had mentioned to me during my fieldwork, both Murthy and Nilekani had keen interest in reforming the state and having India embark on a new path as a nation worth reckoning within geopolitics. Nilekani contested the 2014 election from the Bangalore South constituency on a Congress ticket but lost to his opponent, BJP candidate Ananth Kumar. In his speech, while conceding defeat, he pointed out

---

19. I am grateful to one of the anonymous reviewers for suggesting this apposite term.

20. "Uncertain Times Ahead for Indian IT Sector," *The Hindu*, available at http://www.the hindu.com/todays-paper/tp-business/uncertain-times-ahead-for-indian-it-sector/article3440366 .ece (accessed October 2014).

21. Available at http://www.ibtimes.com/indias-it-outsourcing-industry-faces-strain-global -economic-downturn-853539 (accessed October 29, 2012).

22. Available at http://online.wsj.com/article/SB10001424052702304765304577478193970187 310.html (accessed October 29, 2012).

that "I am disappointed . . . This is my first election. I've learnt a lot in this election . . . I will not quit politics. I will continue working for the Congress. I will not only work for issues related to my constituency, but will continue working for Bangalore."[23] His intention of engaging in public governance is not merely a matter of an election speech but points to his long-standing involvement with the state.

In 2009 Nilekani stepped aside (not down) from his official corporate position in Infosys to serve as the chairman of the new Unique Identification Authority of India (UIDAI). The UIDAI would implement the Aadhaar Scheme, to endow citizens with a unique identification number and also maintain a database of citizens' biometrics and other biographical data. The idea was originally proposed by Infosys as a governance tool, particularly by Nilekani, and it is thus not surprising that he was supervising the initiative. The Aadhaar number is a tracking device similar to the social security number in the United States, but the official narrative states that it is meant to address election fraud and widespread embezzlement of funds that are especially targeted toward poverty alleviation. Tracking populations, in a Foucauldian framework, is a mark of governmentality, and inducting an entrepreneur to administer the system is indicative of a deeper change in the notion of belonging and citizenship, something I argue in this book. Nilekani's role in the state now reiterates my central thesis to understand the extent of the IT success narrative beyond the domain of the industry. However, in 2014 Nilekani stepped down as the chairman of the UIDAI to run for office in the nationwide elections.

Nevertheless, Nilekani's involvement was not unprecedented. The involvement of IT entrepreneurs in public governance has been an emerging social practice for nearly two decades now. For instance, the construction of the new airport was a major topic of discussion at Infosys while I was doing fieldwork mainly because Narayan Murthy was its chairman at that time. The old Hindustan Aeronautics Limited, or HAL, airport,[24] which I used during my fieldwork, was located in the southeastern part of the city. Being of modest scope, it was able to handle a decent volume of air traffic. However, the new airport, which would be built outside the city, was intended to expand the capacity of air traffic in and out of Bangalore. It was promoted as a major stride in transforming Bangalore into a global city, and companies such as Larsen and Toubro, Siemens, and Unique Zurich Airport were consulted for the airport design. In 2005 Narayana Murthy resigned as the chairman after being criticized by H. D. Deve Gowda (former prime minister and then coalition

23. Available at http://timesofindia.indiatimes.com/news/Election-results-2014-I-concede-defeat-to-Ananth-Nandan-Nilekani-says/articleshow/35227480.cms (accessed September 15, 2014).

24. Since 2008 the old airport has been used by HAL and the Indian air force for testing purposes only.

leader of Janata Dal–Secular [JDS] with the Dharam Singh–led Congress) for the delay in the construction and also on issues of land. The controversy over land allocation for the airport brought out issues that were related to land use in general for urban development. The Joint House Committee investigated the airport where Murthy, along with others involved, has been accused of a "poor quality of workmanship" and the absence of an ambience that reflects the specific "culture and glory of Karnataka" (*Economic Times*, December 22, 2009). Even the discourse that framed his resignation is quite revealing.

In a letter Murthy wrote to then chief minister Dharam Singh he mentioned his angst that the time and effort he had put in for five years were not acknowledged.[25] What he did not reference in the letter or his press statements was the related allegation JDS brought about Infosys's application for 845 acres of land in the peri-urban area of the city to expand its facilities. JDS wanted an investigation into how land was being distributed to IT companies and, based on that, whether Murthy was ethically qualified to represent the private wing of a public-private infrastructural project.[26] At Infosys, Murthy's resignation was construed as "unfortunate" not because it was seen as his loss, but because of the state's inability to retain and appreciate, as one engineer affectively put it, "an ethical and globally renowned entrepreneur who was willing to go beyond his duties as a citizen to end corruption and serve the nation."

When I mentioned the land issue to the engineers, it was once again cast as the state's reluctance to appreciate IT's contribution to the nation. Murthy's chairmanship, as the IT industry justified, aligned well with his "mentor" position to oversee innovative work both in Infosys as well as in public infrastructure. In terms of expertise, it is quite difficult to see the connection between the two—software and airport—other than that they are both based on neoliberal tenets. It was, however, known that since the airport would be a critical node in global capitalism, Narayan Murthy was the obvious choice considering his presence in worldwide business. However, Narendra Pani (2006) reminds us that the new airport was just one of many instances of the neoliberal urban reform agenda in Bangalore. The S. M. Krishna government, Pani argues, would identify a set of infrastructural projects and then locate private investors for them. This kind of approach is based on arriving at a technical fix rather than addressing collective urban issues. Therefore, land that was obtained as part of the airport project was based on market acquisition in place of equitable price.

S. M. Krishna, especially in founding BATF, aggressively extended public governance to the private sector to reorganize the city. Nilekani, who headed BATF, worked alongside Ramesh Ramanathan, an ex-banker from London

25. *Hindu Business Online*, October 21, 2005.
26. Ibid.

and New York, Devi Shetty, a prominent surgeon, and others to reform the urban infrastructure that would facilitate the flow of global capital. Janaki Nair has aptly described the coalition of IT entrepreneurs and other professionals charged with the new task of renewing Bangalore as the "captains of the new economy." She argues that the focus of reform was "roads rather than public transport, garbage and pollution rather than public housing" (2005: 336). Matthew Gandy (2008) has noted similar middle-class politics, particularly in reference to water supply in Mumbai. Gandy attributes the dismal state of water supply to the slums to the "extent to which the Indian state has been 'captured' by the middle classes so that its political agenda has been consistently diverted from the universal provision of basic services" (2008: 121). Added to this are middle-class "opt-out" strategies, such as bore wells, pumps, and tankers to access water outside the state provisions, and "environmental improvement," that is, to "clean up" the city for the "respectable citizens" (Gandy 2008). While the Hindu national Shiv Sena movement altered the urban landscape of Mumbai, in Bangalore, the middle-class "capture of the state" was especially emboldened by state representatives, such as S. M. Krishna.

The middle-class "capture of the state" has become a constitutive phenomenon in Indian politics visible in affective events such as the hunger strike by Kishan Babu Rao Hazare (better known as Anna Hazare) in April 2011 to institute the Lokpal Bill (anticorruption bill) alongside the rise of the Aam Aadmi Party (AAP; Party of the Ordinary Man) formed in 2012.[27] Both address the question of corruption and both involved overwhelming participation by the urban educated middle class. The middle-class exasperation about state "corruption," the need for "transparency," and the involvement of citizens in public governance has strengthened over the years, with the AAP sweeping the Delhi Legislative Assembly Election in 2013, though they later lost the 2014 general elections. In the meantime, Janaagraha, an organization that proudly claims it is the "citizens' platform," had extended its work beyond Bangalore with Ramesh Ramanathan being named as one of the members of the Jawaharlal Nehru Urban Renewal Mission (JNURM). The organization, among other things, now manages a nationwide initiative titled "I Paid a Bribe," where a person can anonymously report an incidence of corruption. The website offers a visual image of "India's Corruption Monitor," which dynamically calculates the total amount paid in bribes, the number of reports filed, and from which cities. The moral undertone is obvious; citizens and

---

27. Arvind Kejriwal, a former civil servant in the Indian Administrative Services (IAS), was a prominent Right to Information (RTI) activist and a close associate of Kishan Babu Rao Hazare in that initiative. However, they later split over the question of political engagement through the electoral process, and Kejriwal went on to form the AAP. For a detailed discussion of AAP, see Roy (2014). For an analysis of the connection between the Anna Hazare movement and the contemporary middle class, see Sitapati (2011).

state officials who refuse to engage in bribery are adulated as "Bribe Fighters" and "Honest Officers," respectively.[28]

However, one of the issues that remain constant in my work and in the work of other scholars who have examined urban middle-class activism is the dominance of the supposed middle-class value of "honesty" and the preeminence of the class based on education achievements. Roy, in analyzing the "party-building" strategies of AAP, writes, "Sooner or later (party workers would) return to the 'problem' of the slum voter," who, by participating in "vote buying" to mitigate their harsh conditions, are key to their own misery and, more importantly, in undermining the power of democracy (2014: 51). Like the Anna Hazare movement, where corporate employees participated widely, the AAP, whose leadership was often engaged in "management speak," and Janaagraha, as I show through my ethnography, were not only spaces marked by and for the educated middle class; the pedagogical and didactic mission to elevate the urban poor, with little effort to understand their social context, was an unwavering pursuit. Together, these indicate the preeminence of the middle class not simply as a growing consumer class but one that is invested, directly or indirectly, in the "capture of the state," which has mostly emerged from a sense of "minoritization and victimization" (Khandekar 2013) by the Indian state despite being law-abiding and taxpaying "honest" citizens.

Corruption was a dominant theme in the 2014 elections, especially regarding the incumbent Congress-led United Progressive Alliance (UPA). The electoral pundits in the media and the BJP-led NDA (National Democratic Alliance) made recurring references to the coal scam, the 2G Spectrum provision, the commonwealth game fiasco, the Aadarsh Housing scam, and so on. The final victory of the BJP by a majority is therefore often attributed to the BJP's assurance of a corruption-free nation. However, Chhibber, along with his colleagues, has argued, "There is a crucial difference in how voters think about corruption and how political parties and leaders represent this issue in their campaigns. Voters care more about the corruption they encounter on an everyday basis, whereas parties make the issue a spectacle" (Chhibber, Shah, and Verma 2014). The everyday experience of corruption through which notions of the state congeals is well documented in ethnographic work, as well (Gupta 2005). Even when we look at the "I Paid a Bribe" initiative, the focus is not necessarily on larger scams but the mundane, everyday nature of soliciting, offering, or refusing bribes. It is plausible that the everyday experience of corruption may have resonated with the larger anticorruption tenor of the BJP, which may have led to the win. Regardless, the urban middle class is the single most powerful mouthpiece against corruption considering

---

28. Available at http://www.ipaidabribe.com/bribe-trends#gsc.tab=0 (accessed September 16, 2014).

the perceived or real integrity of its members. Narendra Modi proactively attended to this constituency during his campaign. Later in July his government's budget offered substantial concessions in income tax, savings, and housing loans that disproportionately benefited the middle class ("Onwards to the Neo-economy" 2014).

The other related aspect of the 2014 general election was the definite and obsequious penchant for neoliberalism. Modi is supposed to have won by endorsing the "developmental" agenda, drawing on the "Gujarat Model of Development," the state where he was the chief minister.[29] However, scholars like Jean Dreze (2014) have shown that when we consider Human Development Indexes (child well-being, food, shelter, schooling, health care, sanitation, and so on), Gujarat ranks nine out of the twenty in the state list, far behind Tamil Nadu and Kerala in the south. Gujarat, Dreze (2014) contends, is at best in the "middle" and definitely not the "model." Yet the question remains, as Soundaraya Chidambaram (2014) has pointed out: How does the electorate comprehend development—is it only a matter of GDP or are social inequities taken into account? It is my argument, which aligns with the ethnographic evidence that I present here, that the middle-class capture of the state has shifted the politics of governance and belonging in India to privilege public-private partnerships where the state is required to take a progressively minimal role. Returning to the 2014 budget, it has eased the flow of both foreign and domestic capital primarily through the promotion of public-private partnerships.[30] Overall, the contemporary shifts further reinforce the economic and political processes that brought visibility to IT and organizations like Janaagraha.

Finally, with BJP coming to power, a party that has strong ties with Hindu fundamentalist organizations, such as the Rashtriya Swayamsevak Sangha (RSS),[31] dedicated to the promotion of Hindutva, the concern over whether India is now officially entering the "saffron" (color of Hindu nationalism) phase is legitimate. As an ideology, Hindutva was originally formulated by Vinayak Damodar Savarkar in 1923 as a treatise titled (then published anonymously) *Hindutva: Who Is a Hindu*. Hindutva, in Savarkar's exposition, is not necessarily a religious identity; it is a historical confluence of "Hindus, a Nation," a territory on the banks of the Sindhu' and a race united by the "Common Culture," the common *Sanskriti* (civilization). However, this bond between the land and the culture was interrupted by "foreign invaders," starting with the invasion of Muhammad of Ghazni in A.D. 1001, which led to the conversion of a segment of the population to Islam, who then irrevers-

29. Available at http://www.narendramodi.in/the-gujarat-model/ (accessed September 17, 2014).
30. See "Onwards to the Neo-economy" (2014) for a detailed discussion of the 2014 budget.
31. See http://www.rss.org for details about the organization.

ibly ceased to be part of the land (Jaffrelot 2007). Savarkar's "othering" of the Muslims is one of the prime tenets of Hindutva and underscores the fervent narratives and practices of the RSS and the BJP to quite some extent.

Interestingly, Modi was strategically silent about Hindutva during his election campaign, instead highlighting the "development" plan. This has attracted rather than distract scholars about the future of the nation since Modi is a certified RSS activist (Teltumbde 2014). Anand Teltumbde has argued that rather than taking the rigid approach, Modi is embracing a "soft approach towards Hindutva by systematically saffronising [sic] institutions, as he did in Gujarat," citing the instance of appointing Y. Sudershan Rao, a historian who is known to have sought the virtue of the caste system and determine dates of the epic *Mahabharata* (Teltumbde 2014: 11). Teltumbde also points out that Hindutva aligns well with neoliberalism, which he terms "Saffron Neoliberalism." Both are designed to privilege the already privileged, the Hindus and upper classes, respectively (Teltumbde 2014). The link between the urban middle-class Hindus, for instance, in Infosys and Janaagraha, and Hindutva is not the kind espoused by RSS, but it aligns rather well with Modi's approach, which as I mentioned earlier, can be thought of as "Hindutva Lite." During my ethnographic work I seldom encountered overt references to Hindutva, but the underscoring narrative upheld what was often seen as an intrinsic connection between Hinduism and science. The pride in the success of IT among software developers in Infosys or middle-class activism in Janaagraha were anchored to a vision that had little space for the "other," the "other" being Muslims or the lower classes. In a national political context that assembles Hindutva, science, and the neoliberal market in an unprecedented synthesis, the shifting grounds of public governance that I document here have reached a new level of legitimacy.

The ethnographic work for this book was based in Bangalore, but prior to starting the research, I had also considered Hyderabad as a potential site. Though other Indian cities feature in the success of the IT industry, the Bangalore-Hyderabad dyad is by far the most important because of the exclusive initiatives taken by Karnataka and Andhra Pradesh, the respective state governments. Bangalore nonetheless was historically connected to the science and technology imagination of the nation-building era following independence in 1947. Though the development of the IT industry was finally a matter of government policies both at the central and state level and entrepreneurial initiatives, Bangalore was already available as a space marked by infrastructural industries. These industries were established as part of the nationalist project for technological and scientific self-reliance. Before it became the "Silicon Valley of India," it was known as the "science capital" of the country for its public-sector industries such as the Indian Telephone

Industries (1948), Hindustan Machine Tools (1953), Bharat Electronics (1954), Hindustan Aeronautics (1964), Hindustan Copper (1967), and Bharat Earth Movers (1967). These industries, despite the decline of their standing after the advent of IT, are still present in the collective memory, both inside and outside IT, and they offer a preamble to the new industry. The city, as I discuss in Chapter 3, was also the erstwhile hub of multinational companies such as IBM, which left India in 1978 to return again in 1992.

The 2004 election was decisive especially for urban renewal plans in Bangalore. S. M. Krishna, the chief minister of Karnataka and the architect of Bangalore as a global city, lost the election, and agony over his defeat was widespread both in Infosys and Janaagraha. His defeat was seen as a threat and a setback to the neoliberal agenda in Bangalore. When state power is co-opted, compromised, and weakened, it calls for a critical definition of power that not only looks at the "capture" by the middle class but also at the subversions the poor engage in to survive. The middle-class "capture" of the state is mainly justified as a necessary and ethical antidote to "corruption." However, as Gandy has contended, "The term 'corruption' . . . actually masks an array of different practices and needs to be more carefully differentiated" (2008: 117). In the case of water supply, as Gandy also notes, practices that I have noticed range from bribing for prompt connections, additional payment for dedicated water lines (common for gated communities, which mention 24/7 water supply), illegal connections to the main water pipe, and so on. Though I have known these practices to be common both in middle-class neighborhoods and slums, the middle-class engagement with "corruption" is supposedly prompted by the failure of the state, whereas that of the urban poor is seen as downright illegal. Therefore, the task is to understand the practices and ideologies through which middle-class intrusion into governance can be honored as "reformative," while that of the poor is deplored as "delinquent."[32]

The parochial commitment to urban renewal had implications beyond the city; it seriously affected the rural sector. S. M. Krishna's term also coincided with farmers' suicides in Karnataka, along with other states, which is usually linked to his eventual defeat in the 2004 elections. Based on the findings of the Madras Institute of Development Studies, a research organization in Chennai, *The Hindu* published comparative data in an opinion piece titled "Farm Suicides Worsen after 2001: A Study" in 2007.[33] Karnataka ranked as one of the highest in the country for farmer suicides, leading us to recognize the unfortunate agrarian crisis unleashed by the commercialization of agriculture during the 1990s.[34]

---

32. For a discussion on the anthropology of corruption, see Holler and Shore (2005).

33. Available at http://www.hindu.com/2007/11/13/stories/2007111352250900.htm (accessed April 12, 2013).

34. For a detailed analysis of the farmers' suicide, see Vasavi (1999).

Biotechnology is another important sector of the neoliberal plan in Ban-galore. Lawrence Liang has written about farmers' protests against "biotech-nology and the patenting of seeds and life forms." He specifically mentions the one organized by the Karnataka Rajya Raitha Sangha in front of KFC.[35] Vandana Shiva (2008) has argued extensively about farmer suicides in the context of seed monopolies introduced by corporations such as Monsanto. Shiva mentions four states—Maharashtra, Andhra Pradesh, Karnataka, and the Punjab—where especially farmers who grow cotton, a cash crop, suc-cumbed. Citing the International Food Policy Research Institute's discussion paper "Bt-Cotton and Farmers' Suicide in India: Reviewing the Evidence," she argues that it completely misplaced the suicide events as a "long-term phenomenon" when these happened between 1997 and 2007, affecting nearly 160,000 farmers.[36]

Such misplaced narratives are especially troubling in cities like Bangalore where farmer suicides are incommensurable with the aspiring global image. The newspapers often reported Krishna visiting the farmer's family after the suicide, but the priority of his regime clearly abandoned the agrarian crisis. The overall silence around such suicides during my fieldwork was intriguing. The majority of people I talked to about the farmers' suicide perceived these as isolated problems, and very few even made a connection to the Krishna government's lopsided preference to reinforce urban governance.

The collaboration between corporations and the state is not new in India. Other IT companies, which have a longer history with the Indian state in the pre-IT era, such as Tata Consultancy Services (TCS)[37] and Wipro[38] have also dedicated their profits to social causes through programs of corporate social responsibility (CSR). Infosys, like all the other corporations, has its own CSR, known as the Infosys Foundation established in 1996. The founda-

---

35. Available at http://www.t0.or.at/wio/downloads/india/liang.pdf (accessed October 22, 2013).

36. To see the distribution of farmers' suicide between 1995 and 2010, see http://www.the hindu.com/multimedia/archive/00820/Farm_Suicides__All__820602a.pdf and "In 16 Years, Farm Suicides Cross a Quarter Million," *The Hindu*, available at http://www.thehindu.com/opin ion/columns/sainath/in-16-years-farm-suicides-cross-a-quarter-million/article2577635.ece (accessed October 22, 2013).

37. TCS was established in 1968 and is part of the Tata Group, which was founded by Jamsetji Tata in 1868. After independence in 1947, the Tata Group was one of the main industrial con-glomerates that invested significantly in infrastructure projects to participate and promote in the nation-building project.

38. Wipro, which was earlier known as Western India Products Limited, was established in 1945 to manufacture and sell consumer products, especially household products such as Sunflower Vanaspati (cooking) oil, soaps, and so on. The company stepped into the hardware and software IT industry in the 1980s and in the 1990s rose to global prominence along with other companies in Bangalore.

tion, under the chairpersonship of Sudha Murthy, wife of Narayan Murthy, donates 1 percent of the company profit (after tax[39]) to the development of basic amenities in rural areas in Karnataka and neighboring states. Two issues led me away from focusing on CSR. First, I was not interested in isolated acts of altruism or benevolence because they rarely address social inequality as structural violence; rather often these programs perpetuate the status quo. In saying this, I am not suggesting that Infosys's critique of the state questions issues of social injustice; rather it builds its ideology on existing forms of social inequality while attempting to introduce new ones through ownership and private property. However, the ethico-political critique that I am proposing here seeks a deeper political and economic transformation that is not limited to isolated attempts (building schools, hospitals digging wells, raising flood relief, and so on) to "help" the rural poor.

The critique is an attempt to overhaul the governance apparatus in a specific way, that is, to replace the socialist-redistributive model with one based on the neoliberal market. Second, the Infosys-Janaagraha bond for transforming governance was singularly focused on the city, which was seen as an important node in the circuit of global capital. One of Janaagraha's missions was to revitalize urban politics and the urban domain, which it argued had suffered because of state policies that privileged the agrarian economy and the rural sector. My inclusion of the Water Project alongside software illustrates this argument. The historical amnesia is deeply linked to the ethico-political critique because while attempting to reform the nation-state, the Infosys-Janaagraha collaborative erases its past as problematic and burdensome. The claim I make in this book is not about interchanging Infosys with IT and then extending the argument to India as a whole. Instead I think of Infosys as amply reformative based on its corporate power, to be able to initiate changes in the government apparatus not only at the state level but also at the level of the central government. Nevertheless, this could only happen at a certain moment in the history of the nation-state, and my intention is to capture the political nuances of that moment ethnographically. Therefore, this book explores emerging alliances that recalibrate the relationship between state and citizens.

Over the last two decades, the rise of IT in general, and in India specifically, has generated significant interest among scholars across disciplines in the context of globalization, neoliberalism, its implications for transnational labor, and so on. Manuel Castells's (1989) "information society" (informational economy, global economy, network enterprise, flexi-workers, timeless time, annihilation of time, space of flows) comes to mind; he offers

---

39. This was specifically emphasized as I was learning about the Foundation during my fieldwork.

us ways to theorize the restructuring of capitalism transformed primarily by the IT revolution in information processing. Other scholars have focused specifically on the Indian IT industry. For instance, in their edited volume, Carol Upadhya and A. R. Vasavi explore the various facets of the Indian IT industry at the intersection of culture, corporate management, the production of the Indian techie, labor processes, gender, and so on. In *Virtual Migration*, Aneesh Aneesh (2006) offers us a reading of the outsourced model of software development as the decoupling of the "body" from its "performance." It generates, he argues, a different kind of temporal integration when servers, software engineers, and clients are differentially located across incongruous geographies and time zones. Christopher Kelty's (2008) *Two Bits* focuses on free or open-source software and how it leads to the emergence of what he calls a "recursive public." A recursive public is a specific kind of public, composed of "geeks" who are at the forefront of establishing a different kind of public space that contests existing forms of power. In terms of power, free software, which developed rather unexpectedly, counters all that is associated with economic history, especially with capitalism, such as "private ownership" and "individual autonomy." I knew a handful of geeks at Infosys who were committed to open-source software, which was quite in contradiction with the capitalistic mission of the company. Though my research does not focus on the politics of open source, Kelty's work was helpful for me to think about the politics that underscore software development work, such as the sharing of tacit knowledge. Kelty's work, unlike mine and like the other works I mention here, once again is multisited. In mapping a "distributed phenomenon ethnographically," his work was spread over Bangalore, Boston, Berlin, and Houston.

Another area of focus in this literature has been on gender. Men, especially in the upper echelons of companies, dominate the IT industry. This is immediately visible, and the IT industry is marked by a gender inequity that is quite characteristic of the industry itself. The ideal software programmer is a man ideally without a family. If he is married, the wife is expected to be in a supporting role, helping the man climb the corporate ladder (Devi 2002). Women programmers I knew at Infosys often complained about managing long hours of commute, work, domestic responsibilities, and family obligations, especially toward in-laws. A few women also reported that to help them advance professionally, their parents were taking care of their children, who occasionally lived in other cities. Then there was the question of international travel to a client site, which most women were either not offered, had to pass, or had to accept along with the emotional cost of "abandoning" their children. At the time of my fieldwork there was a conversation about opening a daycare for the employees' children at Infosys, but most of those meetings were organized and attended by women. The implication for gender equity in a neoliberal context that purportedly upholds individual pursuits is favorably skewed toward men.

Scholars such as Winifred Poster (2013), Danielle van Jaarsveld and Poster (2013), Reena Patel (2010), and Rupal Oza (2006) analyzed labor relations between the West and India as a function of time zone differences and their implications for questions of personhood, gender, and the circadian rhythm. Smitha Radhakrishnan's *Appropriately Indian: Gender and Culture in a New Transnational Class* "is about the culture of India's IT professionals—a culture that powerfully reconciles the 'global' and the 'Indian' with one another to create a simplified, influential notion of a new Indian culture that is compatible with the economic and geographic mobility of the global economy" (2011: 3). The relationship between gender and neoliberalism has taken another critical turn with reports of violence against women who work late hours in the urban business processing outsourcing (BPO) companies, especially the call centers.[40] This raises an important question about the enduring marginalized place of women in an economy that is otherwise perceived to be enabling individual aspirations. On the other hand, because often the men accused and convicted in these violence cases are also from the lower class and/or rural migrants, the divide between the "progressive" city and the "regressive" village grows exponentially.

Gender's relation to water is an important feature in the literature. While I do not explore this specific connection, the fact that my ethnography on water dwells mostly on women is pertinent. In the face of water scarcity in the slums, women were largely responsible for collecting water for the family, and especially making sure that the men had enough water to use. This ranged from walking several miles to a public fountain to placing pots and buckets in the queue when water was released for the slum. Further, when we look into developmental projects like the Water Project, gender takes on a different character, an issue donor agency personnel often overlooked.[41] During the meetings the Water Board held in the slums to explain the advantages of the Water Project, the voices and opinions of the male residents were most active and effective. Therefore, the chasm between the water chores and the power and decision making followed a strict gendered line.[42]

Caste is another subject that scholars have explored in their examination of the IT industry. In my ethnography it was clear that the IT industry positions itself as an open forum, where success is based purely on achievement. Unlike traditional sectors where ascriptive status matters, IT is perceived to be based solely on "merit." However, the constitution of founders and employees at Infosys (and other companies) is significantly dominated by the upper

40. The Associated Chambers of Commerce and Industry in India released a report in January 2013 highlighting declining female work productivity in the BPO sector. Available at http://www.assocham.org/prels/shownews.php?id=3843 (accessed October 20, 2014).
41. See Coles and Wallace (2005).
42. For the politics of gender in the case of water, see Lahiri-Dutt (2006) and Coles and Wallace (2005).

castes (and upper class). In her study of software professionals in Bangalore, which examined educational opportunities available to upper-caste (and upper-class) urban citizens and recruitment strategies of IT companies, Carol Upadhya (2004) showed that there is a critical gap between the IT rhetoric of "merit" and the actual reality.[43] The exclusionary practices of IT has certainly expanded the middle caste-class beyond the traditional professions (professors, lawyers, doctors, engineers, and so on) but in the process has also deepened social inequalities.[44]

While the above literature has shaped my thinking, my own query is somewhat different. What continues to be either explicit or implicit in the research is the idea that globalization exists as an intangible network that recruits the nation-state as its appendage. I am not claiming that the network is prior to the nation-state or vice versa, where scholars also debate over the demise or resurgence of state sovereignty. This is not a matter of succession, but of how the global network and the nation-state influence governance practices in the name of neoliberal reforms. In other words, how does one experience, endorse, or resist neoliberal changes in everyday life? My ethnography on the creation of the water market is meant to trace the processes through which people variously encounter its change from a basic amenity to a commodity.

Saskia Sassen (2005) argues that the capacities for global operation, coordination, and control contained in the new information technologies and in the power of transnational corporations need to be actualized. I extend Sassen's formulation to further argue that this embeddedness is not merely a question of actualizing place but also an invitation to historiography. In this sense the practices, discourses, and processes of neoliberal global capital cannot be detached from, to reference Michel Foucault, the "history of the present." The "history of the present" of IT is not about the historical background but how the past gets written into the present. Therefore, distinct from the scholars I am in conversation with, I attend to the question of globalization, neoliberalization, and IT by visiting the postcolonial historiography. In fact, I argue that the postcolonial condition constitutes, deploys, and resists global forces in a way that is reminiscent of an earlier imperial past, a travail we cannot ignore.

My interest in IT does not exclusively focus on its global network of flows of capital, software, and programmers. Rather, I am interested in its politics within the nation-state and how it bears on the issue of citizenship and belonging. Can this work be considered multisited? Though I remained in Bangalore, I did travel through disjunctive and incongruous spaces in the city. For example, a typical day in Bangalore involved a negotiation between

---

43. For further discussion on how education plays a role in IT, see Krishna and Brihmadesam (2006) and Fuller and Narasimhan (2006, 2010).

44. For a caste-class continuum, see Fuller and Narasimhan (2014).

the filth-logged storm drains in the slums and the opulent spaces of IT corporations. Ethnography in such incommensurable spaces may be termed multitemporal because, in the IT discourse, it represents a linear progression from "backwardness" to "modernity." Contrary to its promises, neither globalization nor neoliberalization has generated an even political terrain. By political terrain I mean the global as well as the national terrain. This work is an attempt to map the ruggedness of this uneven terrain as a way to show how the IT narrative differentially affects everyday lives of hierarchically placed populations. Ethnography is a particularly useful method in this enterprise because it compels the researcher to follow a story not for its cohesiveness but for its abundant missing pieces.

The book is organized based on the following chapters. Along with an outline of the manuscript, the first chapter, "The Politics of Distrust," introduces the various sites *in and through* which I came to formulate the project, moving back and forth spatially among them rather than offering a chronology of events. Methodologically, these sites intertwine in terms of their shared investment in thinking about "India," "Bangalore," "citizenship," and "governance." Here I propose a heuristic binary tool—"inside" and "outside" IT—to show how these notions are constructed and contested along class lines. The inside of IT is the domain of software development, where work and human processes intertwine to meet the stipulations of the global economy. The outside, however, is not merely what lies outside IT or is excluded from it. Rather, I deploy the "outside" as a way to reverse our gaze on IT. Therefore, instead of looking outside from inside IT, the outside offers a critical analysis when we look into IT. The outside is where the Water Project is located, but it is also a multilayered and contested domain where different classes lay their claims to the city. The inside-outside is not a dichotomy that is sealed; it is the new porosity between the two otherwise unrelated domains, that is, software and water. These now converge in the neoliberal discourse and offer us a way to revisit the question of citizenship and belonging.

Chapter 2, "Market Experiments and Peripheral Lives," discusses the basis of *why* I bring the two domains—water and software—under the same rubric of analysis. It explains *how* these domains become commensurate in thinking about citizenship and governance. I provide the details of the Water Project and explain how it is considered unprecedented in India in the area of public infrastructure work. For example, the citizens' contribution toward the capital cost and their participation in monitoring the project as one of the stakeholders are two of its unique features. It also discusses WSP's interest in funding the "citizen participation" module and International Finance Corporation's insistence on privatizing the water supply. Interestingly, though the Water Board was administering the Water Project, the territory that was included, that is, the peripheral region circling the city of Bangalore known as Greater Bangalore (the area was later merged with the city), was technically beyond the limits of Bangalore Mahanagar Palike. I specifically highlight this peripheral region as an "available" space for neoliberal experiments. I show the parallels that emerge between a software/corporate project and a state/public project where one's participation is circumscribed only as a worthwhile "stakeholder" in the market.

"Software Nation," Chapter 3, explores the internal processes of the IT industry, focusing on several different units in Infosys that handle different segments of the job. It would be safe to consider this chapter as the ethnography of IT. However, I extend this ethnography to decipher a "blueprint" of governance and citizenship that is possibly emanating from within this domain. Through this ethnographic account, one encounters not only desirable and successful citizens but also ethical corporate governance practices that Infosys claims it closely adheres to. Here I mainly focus on the following set of questions: What are possible indicators through which we can recognize this blueprint, as Raymond Williams suggests, as "emergent" and "residual"? How are these new indicators constructed as a neutral body of governance tools that can seamlessly alternate between corporate and the public governance? What makes such alternations possible? Why now?

I explore one aspect of the success narrative that I encountered regularly at Infosys—"We put India on the map"—in Chapter 4, "Travails of Time." This segment of the narrative is ahistorical in the sense that it frames IT as a sudden occurrence that not only marks India's departure from a troublesome colonial past but is also delinked from the "corrupt" postcolonial present. Chapter 4 interrogates this notion of departure and delinking. I show how despite all its claims, IT in fact folds in what we may call a "history of the present" through a selective borrowing and denial of the past. Issues of belonging and citizenship are underscored by contestations through time; therefore, this chapter attempts to anchor IT in historiography, not to provide a historical background but to understand the changing terrain of the nation-state with the emergence of neoliberalism. I pay special attention to the reemergence of Gandhi as the new referent and the simultaneous demise of Nehru in the IT narrative to attend to the question of history.

Chapter 5, "The Black Box," deliberates on how although the neoliberal reform agenda privileges the upper classes given their economic strength, the urban poor were omnipresent. Their presence had little to do with inclusion but was a question in itself: What is to be done with the poor? The poor were clearly "residual" yet persistent in disrupting the neatness of the neoliberal picture. Drawing on my ethnographic work in a slum in Bangalore, Manjunathanagara, which was part of the Water Project, the urban poor question was critical in two ways. First, Janaagraha devoted a significant amount of time to what it called the "urban poor question," basically trying to incorporate them within a framework that otherwise was designed for middle-class citizens to participate in the Water Project. Second, the urban poor themselves were resisting the payment that would be incumbent if water supply was privatized. Related to this, they would lose a vote-bargaining strategy with their local councilors. Through this chapter I show that the experience of neoliberalism is far from being a monolith and that the market is not necessarily a tool favorable for allocation of resources since it tends to homogenize the polity despite its clear discontinuities. As I mention in Chapter 1, the urban poor on the "outside" of IT, in fact, lend us a different looking glass through which we consider the politics of the market reform. This chapter reverses our gaze on IT. It compels us to pause and think about how market reforms are reinforcing existing forms of inequality while instituting new ones. Finally, in the conclusion, "BITS of Belonging," I revisit the question of postcolonialness and the restlessness that marks its continued departure from a past that is degrading. The success narrative of IT is one such instance that attaches itself to the neoliberal ideology to argue for the renewal of the nation-state. Yet, as I show, this renewal is limited in terms of who can now belong as a substantive citizen—that now one can only belong in bits.

I call this book *BITS of Belonging* to tell a complex story of postliberalization India, particularly how the disenfranchised population copes with the changing political paradigm and devises new strategies to make claims to the nation-state. This claim, as I see through my work, is a fractured reality—a binary of 0 and 1—as in a computer BIT. Neither 0 nor 1 matter as entities; the various combinations of the two constitute a legitimate software program. The combination has a predetermined and limited field of possibilities. Likewise, belonging ceases to have an amorphous stance and becomes a definite combination of politics, ideology, and economics to be recognizable as a legitimate way to exist in the nation-state. Yet the urban poor contest this form of citizenship as a way to continue their claim on the state. In this sense citizenship and governance are both a venue for reformation as well as a contestation. The success story of IT, despite its reassuring tone, serendipitously becomes a conduit to class politics that reflects that liberalization, or globalization, for that matter, is far from a panacea.

1

The Politics of Distrust

The IT industry in India rose to global prominence in the last two decades as the center for software application development. Since its inception, the industry has responded to the outsourcing business model adopted by corporations mostly in the United States, Western Europe, and Australia. In this model, the corporations do not use their own resources to address their software requirements but subcontract them to IT companies in India noted for their excellence in software programming. As this model also involves relegating jobs across continents, it is also known as the offshore-outsourcing model. This model primarily hinges on a detailed negotiation between the client's enunciation of its business needs and the IT company's ability to translate those needs into a software package using various computer languages. Thus, given the primacy of the outsourcing model, the Indian IT industry has a negligible market presence within the country, making it an exclusively global industry.

Though India has long traded with the West, the success of the IT industry is described and celebrated as unprecedented because IT, a high-tech feat, occurred in India despite its standing as a developing country. The NASSCOM projects that IT exports will touch close to $100 billion in FY 2015.[1]

As we may recall from the previous chapter that the industry's contribution to the GDP is relatively low, though NASSCOM casts this contribution

---

1. Available at http://www.nasscom.in/india-itbpm-exports (accessed April 8, 2015).

in terms of "highest relative share in the national GDP."[2] Yet the IT narrative often and rather triumphantly talks about how the global presence of the industry is a boon to the national economy on the whole. Let us look at NASSCOM's explanation:

> [The] IT-BPO sector has become one of the most significant growth catalysts for the Indian economy. In addition to fuelling India's economy, this industry is also positively influencing the lives of its people through an active direct and indirect contribution to the various socio-economic parameters such as employment, standard of living and diversity among others. *The industry has played a significant role in transforming India's image from a slow moving bureaucratic economy to a land of innovative entrepreneurs and a global player in providing world class technology solutions and business services.* The industry has helped India transform from a rural and agriculture-based economy to a knowledge based economy.[3]

While economics is a vital part of the IT narrative, it is not exhaustive. Instead, there is a more important and impassioned rhetoric that states the transformative effect IT has had on the image of the nation as an emerging global player. Stressing this national emergence alongside the departure from an agriculture-based economy showcases the advanced technical knowledge base that the country has to offer. NASSCOM report also states that the sector provides jobs directly to 3.5 million people,[4] which in a country of close to 1.2 billion people is miniscule. Despite such numeric disparities, IT is upheld as a success both inside and outside the nation.

Consider the following response from Nandan Nilekani, the chief executive officer of Infosys Technologies, when I asked him how he feels about the success of the IT industry in India:

> Certainly it [his work in IT] is exciting. IT industry is pivotal in terms of creating an outlet for aspirations of young people. It is pivotal in creating a niche for India in the global economy, for generating revenues and growth; IT is also an instrument for social development. There are so many dimensions to this that it is fascinating . . . this has such a multiplier effect of what we are doing and particularly for those of us who are saying what we need to do to make India a better place to live, free of corruption and comprised of free citizens.

---

2. Available at http://www.nasscom.in/impact-indias-growth (accessed April 8, 2015).
3. Available at http://www.nasscom.org/impact-indias-growth (accessed April 5, 2013). Emphasis mine.
4. Available at http://www.nasscom.in/impact-indias-growth (accessed April 8, 2015).

As one can see, the narrative hardly focuses on IT as a specific system of knowledge and India as one of the major players in the current world economy. Rather, it dwells on the kind of transformative energy IT's success can release for the nation-state. The "multiplier effect," as I elaborate in a later section, is Nilekani's prescription for social change and national renewal, where the Indian state, specifically politicians, is reprimanded for corruption and failure. IT professionals often argue that failure is largely the result of the socialist-redistributive model adapted by India since its independence in 1947, which does not acknowledge citizens as rightful *owners* of the nation and/or their individual freedom. In the IT narrative, the Indian state is often seen as a replica of the previous colonial state, which produced subjects of rule rather than citizens. On the other hand, citizens are encouraged to participate actively in democracy as part of their stake in the nation. The proposal mainly encourages the state to adopt the model of corporate governance and its standards—"transparency" and "accountability"—for public governance. These standards, which are the core processes of the neoliberal market, are tools to supposedly eliminate barriers between the service provider and the client to ensure the free flow of information between the two.

In other words, this is a change that endorses the "openness" of the neoliberal market as to expose the "closed" state. For some time now, developing countries have embraced the market as their last effort to overcome economic hardship generally stipulated by the international donor agencies like the World Bank. India is not different in this regard. The country has a long history of development projects and structural adjustment programs as collateral for international loans. However, the success of IT to some extent detaches India from the stigma of a developing country and places it closer to the developed world. IT thus provides the opportunity to connect with the global circuit to talk about "freedom," now more than ever espoused by the developed world. Thus, IT in its adjacency with the West both elevates and alleviates India, promising to offer a direction of national revival that will transform the nation-state to align with Western democracies.

In almost all of my conversations, IT professionals were particularly gratified to be at the forefront of this significant national emergence. Yet seldom did any of these discussions close with the recounting of "success" only. Rather, success was, at best, the frontier of the IT narrative. IT was impatient to flow beyond its corporate confines—whether in actively engaging with programs of social change or informal discussions among colleagues— to the wider social domain. By impatient I mean that there was a palpable sense within the IT community that India is deteriorating and there is an urgency to retrieve the nation-state. The success of IT in this frame offers a historically opportune moment that India should seize to rise beyond its depleted condition and at the same time distinguish itself from other developing countries.

The artifice of inclusion is central to the IT narrative palpable, for instance, in books such as *A Better India: A Better World* and *Imagining India*, written by Murthy and Nilekani, respectively.[5] Murthy once passionately wished that "India should reach a level where every child has enough to eat, proper healthcare, a roof on [sic] head and has access to education."[6] However, a closer examination of the market ideal, that is, consumerism-based citizenship, reveals that it does not, or rather cannot, address issues of social deprivation. In fact, the market begins by strengthening already existing class privilege while also creating new forms. Its exclusionary nature forms the very rationale of its competitive ethos of freedom so desired by consumers. As a standard capitalist calculation of supply and demand, the market is also a vital tool to suppress dissent. The de-recognition of dissent increasingly pushes a society toward a condition of depoliticization or antipolitics. Politics and market are thus perceived as antithetical, as though they are detachable in reality.

Undoubtedly, the apparently inclusive narrative is sheer tokenism toward the poor. However, to push this argument a little further, I argue that the token inclusivity serves a specific purpose; that is, in this changing global capitalist scenario, poverty and the urban poor are another commodity. The commodification of poverty does not concern actual people or their lived experiences. Rather, it is the condition of poverty—nameless, peopleless, and motionless—that is harvested for demonstrating democratic practices, such as participation. Conversely, as I show later, there is an emergent reluctance on the part of the urban poor to participate, which we may want to think of as a politics of refrain. In this sense, the IT narrative inspires the disenfranchised to follow new options for resistance despite the market's attempt to depoliticize society.

What would the corporatization of an existing socialist redistributive state involve both epistemologically and ontologically? The last two questions are crucial in the sense that my engagement with critical anthropology took me along paths remote from IT—to the slums of Bangalore. The slums were not just non-IT, remote from it; rather, and more importantly, they were un-IT. Success stories or "happy" stories are always problematic because in their urgency to celebrate, they conceal more than they reveal. Happiness at the same time is also an assertion of power in that it is able to render the "unhappy" trivial in its narration. Billboards in the city broadcasting "Happiness Sale" for department stores or "Make Way for Happiness" for new car models were ubiquitous. Though the happiness quotient has now extended to other cities with neoliberal aspirations, such as Gurgaon in the northern state of Haryana, it undoubtedly premiered in Bangalore.

---

5. See Nandan Nilekani, "Ideas for India's Future," available at http://www.ted.com/talks/ nandan_nilekani_s_ideas_for_india_s_future/transcript?language=en (accessed October 2014).

6. Available at http://www.mid-day.com/news/2004/apr/80987.htm.

Yet the "unhappy" refuse to disappear; they continue and struggle to thrive in derelict spaces, both social and physical. The slums as the "unhappy" abandoned spaces contradict and threaten the "happiness" of IT. To underscore the continuity of the poor and the underprivileged for a city poised at the brink of flight as an important node in global capital, this ethnography focuses on those who exist in the margins of the neoliberal market. My examination of the Water Project is a way to understand how the urban poor coped with the new "happy" consumerist rhetoric that had real consequences. First, it involved a contribution from citizens who were "stakeholders"; second, a memorandum of understanding (MoU) was signed between Janaagraha and the government of Karnataka for citizens to ensure "transparency" and "accountability" in the project on May 21, 2005.[7] Finally, once the project would be completed, the operation and maintenance (O&M) phase would be privatized.[8]

The reorganization of water, a basic necessity, as a commodity offers a productive way to see how it overlaps with the discourse of capital goods, such as software. By bringing these two otherwise unrelated domains—water and software—within the same analytic frame, I explore the politics that underscore neoliberal ambitions of the middle class and the everyday resistances they generate among the disenfranchised populations in the slums. These various and apparently incongruous sites, on the other hand, show the limits of the success discourse. I indicate two related issues: First, not everybody in India can be part of this successful industry. Second, the success discourse, which aspires to lift India, will in effect lift only the middle class since IT is an exclusive middle-class success story, and, not surprisingly, Janaagraha is also concerned with middle-class issues.

I find Raymond Williams's (1977) dialectic triad—the emergent, the residual, and the archaic—a useful way to understand the politics between the privileged and the underprivileged. Williams writes, "By 'residual' I mean something different from 'archaic.' . . . I would call the 'archaic' that which is wholly recognized as an element of the past. . . . The residual, by definition, has been effectively formed in the past, but it is still active in the cultural process, not only and often not at all as an element of the past, but as en effective element of the present" (1977: 122). The urban poor and the continuity of poverty in cities like Bangalore that aspire to be globally recognized continue as the "residual" since their presence and their politics frame the urban space. The urban poor are by no means archaic; they cannot be relegated to the past.

---

7. Janaagraha later withdrew from the MoU in 2006, stating that they were unhappy about the transparency processes followed in the Water Project. However, as I show along with other critics, the decision was primarily in response to the criticism of Janaagraha's participation in the privatization of water in the city. Though the question of the urban poor was often discussed in meetings, it was usually from a middle-class perspective, as I demonstrate in Chapter 5.

8. See Walters (2013) for the wider initiative to privatize water in the state of Karnataka.

The emergent, on the other hand, as Williams argues, is that which lends itself to "new meanings and values, new practices, new relationships and kinds of relationships are continually being created" (123).

On my first visit to Infosys in 2002, I talked to a diverse group, including the founders, Nandan Nilekani, members of management, software architects, project leaders, programmers, the knowledge management team, human resources, and others. Over the various conversations, it soon became evident that Infosys occupies a special place in the minds of its founders and employees not only as a successful company in the global IT business but also as the epicenter of a nation poised for phenomenal social change. "What does working in the IT industry mean to you?" I consistently posed this question to everybody I spoke with in order to fathom the depth of the success narrative among people who were at its center.

Though most of them were excited to be part of a cutting-edge technology, the high remuneration IT companies offer was often cited as an important reason to join the industry. Some informants candidly related their profession to their "happiness" and "well-being." Specifically they stressed their financial power and the ability to participate in newly introduced lifestyles, such as flats in gated residences, imported cars, high-end gadgets, overseas vacations, expensive private education for the children, regular visits to malls, dining at fine restaurants, and so on. To these others added the climate-controlled work ambience IT offices offer versus sweltering factory shop floors. Others were excited by the prospect of spending some time at a client site in the United States or Europe as part of a project team to experience the "West," which was valuable as both a résumé builder and a means to acquire social status.

Despite all its variations, the response to my question inevitably cascaded into *how* IT is instrumental for overhauling the image of India in the world. The only difference I noticed was that while the founders and management interpreted my question to be about the condition of India, others took a brief detour before they started talking to me about the country. The swift transition from IT as an industry to IT as a medium for national reformation points to a sense of impatience with the existing system of public governance. That IT was not limited within IT but was the national vanguard became increasingly clear.

The partition and the zones that are instrumental in producing, animating, and managing the "new" nation, mainly in the form of spatial, material, and virtual restrictions, are significant to the analysis offered in this book. Through this work I speak to the limits of corporate ethnography: the difficulty of defying corporative performance expectations and rituals. For example, while I was allowed access to all facilities at Infosys, like their employees, my movement as a researcher was channeled through the rituals of surveillance and checkpoints, which demanded specific documents, such as the identity cards/visitor pass (in my case), screening of bags, and logging of work hours.

A similar corporate process was evident in Janaagraha, as well, though it was not digitally controlled to the extent it was in Infosys. Rather, the work ethic was enforced through the moral rhetoric of diligence, honesty, and punctuality—retrieving national values that they considered lost. Besides, Janaagraha also resisted being seen as just another NGO since their work focused on a fundamental change of governance and citizenship rather than isolated social initiatives. They preferred to think of their organization as a citizens' "platform for change." To this extent, the work was relatively more structured, with weekly reporting procedures along the hierarchy as in a corporation, which ensured both transparency and accountability in their work. For example, an organizationwide meeting was held every Monday morning at nine o'clock. This was followed by different groups meeting individually throughout the week to report on their progress over the past week and present agendas for the current week. Since the founder and the prominent members shared a corporate background, the model was not far from an obvious choice. For instance, the work done by Janaagraha on individual projects was always calculated and presented to the public in terms of "man-hours."

Interestingly, IT companies also bill their clients according to the number of man-hours spent on a project. These kinds of measurement strategies were aimed at the corporatization of civil society and mark a departure from traditional social work in India, which was relatively more amorphous. Every project in Janaagraha was directed toward augmenting citizen participation without much attention to existing social inequalities that may hinder such participation, especially of the urban poor in a space and rhetoric dominated by middle-class priorities. I show that these procedures, which are endorsed as "scientific" and "computable," were critical for the reemergence of the middle class and the institution of the market. My visit to other NGOs in the city, which operated in a relatively unstructured manner, where responses and decisions were more fluid, confirmed the new mode of civil society organizations Janaagraha was keen to introduce.

What intrigued me at length both at Janaagraha and Infosys was an ambivalent politics on the issue of social inequality. Since most conversations dwelled on the unfortunate condition of the nation, I wanted to know their thoughts on poverty vis-à-vis their middle-class location. There was a quick acknowledgment of poverty, but the unwillingness to engage with the lived condition of the poor soon buried the issue. Poverty here offers a particular insight into the middle-class engagement with the question of class difference. On one hand, the acknowledgment is a way to build distance between middle-class affluence and the dereliction of the poor. On the other hand, it is now marketable in global capitalism not only as a commodity but also as a necessary collateral of democracy. In a nation that since independence has actively responded to issues of social inequalities as matters of politics, democracy, and policy, both successfully and unsuccessfully, such studied distances are new and troubling. Further, since the middle class was replicating itself as the vanguard of change along the lines of the nationalist struggle, the rhetoric of social distance raised some serious questions.

The middle-class capture of the state was overtly and confidently justified in Janaagraha on the basis of education and professional achievements. The urban poor, it was often argued, had little idea of democracy and the market, of their functioning, expectations, and transformative power. Education, which has long been a class separator in India, was ritually used to carve out a space for middle-class politics. By this logic, the middle class understood and appreciated the rationale of the market as the principle of governance, the complexity of globalization, and the urgency of national reform vis-à-vis the ignorance and irrationalities of the urban poor. The veneer of education also disguised the otherwise absurd assumptions about the market the middle class espoused, most troublingly, that the market is a politically neutral tool. However, the urban poor had to be educated and eventually inducted if the market were to operate seamlessly. The slum as a space peripheral to Infosys's and Janaagraha's aspirations to transform the nation-state defies the universalizing thrust of the market paradigm. In another sense, while IT exalts life and productivity, the slum is a space, adjacent yet obscured, that exudes the opposite: the violation of basic human existence.

The ethnographic details of the everyday struggle for survival around basic amenities, especially water, are significant: dependency on the valveman to turn on the supply, anticipating the time, tolerating inconvenient hours of water supply, waiting in long queues to collect water, amassing as much as one is allocated for that particular day, efficiently storing that water, carefully rationing its use, and finally dealing with its diminishing volume over a week when the wait begins again. The "outside" is a move to turn the lens onto to IT itself. Far from being a neutral tool for social change, the "outside" reveals the inherent political and ethical fractures of the market. Through various ethnographic vignettes, I show how the urban poor actively pursued the pertinent questions about the water market that were never raised in the Infosys-Janaagraha circuit. The "outside" also reinstates my critical location as an anthropologist, as a way to find my own voice interrogating the corporate discourse and its everyday rituals. The juxtaposition of these rather disjointed everyday experiences—the global ambition of Infosys and

Janaagraha and the everyday survival struggle of the slums—reveals a deeper social disconnect that otherwise is easy to gloss over, especially in a city identified as the neoliberal core of a developing country.

The Watershed

Though authors are divided over the beginnings of economic liberalization, 1991 is generally considered a watershed year in contemporary India.[9] That year, facing a severe fiscal deficit, the Narasimha Rao (then prime minister) government, under the aegis of the then finance minister Manmohan Singh, launched market-oriented reforms that departed from the socialist-redistributive and state-regulated model that had been in place since independence in 1947. Commercially, they opened India's economy to foreign investments and also removed myriad internal bureaucratic constraints such as licensing and taxation to help Indian companies trade more easily in the global market.[10] This policy transition was designed to exploit the competitive advantage of the Indian IT industry in the emergent global market, which would then trickle down as general economic growth for the country.

Understandably, the industry that plausibly benefited the most from these reforms was IT and other related sectors. Since the early 1990s, IT has established a robust global trade in software development work outsourced mostly from businesses in the United States and Western Europe (Grieco 1984; Moore 2000).[11] So consequential were these reforms that Narayan Murthy likened them to the "winning of economic freedom, on the lines of the securing of political freedom from British rule in 1947."[12] He further remarked that the policies "changed the Indian business context from one of state-centered, control orientation to a free, open market orientation, at least for hi-tech

---

9. Some authors argue that the process of liberalization had started in India earlier than 1991 with the drafting of the Computer Policy in 1984 and the Computer Software Export, Development, and Training Policy 1986 under then prime minister Rajiv Gandhi, which is sometimes also referred to as "halting liberalization." For more details, see Parthasarathy (2004). Still others locate the start of economic reforms as early as 1973, when isolated attempts were made to ease the over-regulated economy. See Girdner (1987) and Hardgrave and Kochanek (1986).

10. In their edited collection, *India in the Era of Economic Reforms*, Sachs, Varshney, and Bajpai focus on the economic crisis of 1990–1991 when "the gross fiscal deficit of the government reached 8.4 per cent of GDP, and the annual rate of inflation peaked at nearly 17 per cent . . . this near miss with a serious balance-of-payment crisis was the proximate cause that India's market liberalization measures in 1991" (1999: 1–2).

11. Grieco (1984) analyzes the IT industry in India before the 1991 reforms in the context of a developing country vis-à-vis its relationship with powerful multinational companies who then were the dominant IT companies in the country.

12. "Reflection of an Entrepreneur," Narayan Murthy's address at Wharton Business School, Class of 2001, available at http://64.233.169.104/search?q=cache:TWtSbXnLmOOJ:www.infosys .com/media/Wharton_Reflections_May01.pdf+winning+of+economic+freedom+murthy&hl= en&ct=clnk&cd=1&gl=us (accessed March 2007).

companies."[13] An *Economic and Political Weekly* editorial notes, "The sterling performance of Indian software professionals and entrepreneurs in Silicon Valley has convinced the world that India and Indians have an edge over the rest when it comes to IT. Consequently, developed country capitals have, for the first time, started viewing India as a potential partner" (*EPW* 2001).[14]

In one of my conversations with him, Murthy asserted that the "wealth we created is different; we did not pay anybody any bribes and neither will we accept any." He referred to several incidents where Infosys did not compromise in paying bribes to state officials even when the business could have been jeopardized. One of the key corporate values of Infosys, "The softest pillow is a clear conscience."[15] Murthy explained that Infosys arrived at this maxim "after extended deliberation as how best to convey our contempt for corruption because it is important for people outside as well as for our employees." However, since the time of my fieldwork, the IT industry has been hit with some ignominies of financial fraud, the most prominent of them concerning Satyam Computers in 2009. The following is a report from *Time* magazine on what has come to be known as the "Satyam Scandal":

> According to a 200-page CBI report, Satyam insiders forged board resolutions to secure $260 million in bank loans which were diverted for personal use, and over several years generated fake customer identities and account statements to inflate Satyam's revenues by millions of dollars, boosting the company's share price and making its books look far healthier than they were. Investigators following the paper trail have discovered that embezzled funds were channeled into 1,065 properties valued at $74 million, including some 6,000 acres of land, 40,000 sq. yd. of housing plots, and 90,000 sq. ft. of other developed real estate. The properties, bought in the name of some 80 shell companies, included prime commercial plots in and around Hyderabad, Bangalore, Chennai and Nagpur.[16]

Following the scandal, the Planning Commission deputy chairman, Montek Singh Ahluwalia, assured the world in the 2009 World Economic Forum in Davos that "as long as we handle it properly and it is seen that wrongdoers

---

13. Ibid.

14. NASSCOM surmises the merit of the 1991 reforms as: "The liberalization and deregulation initiatives taken by the Indian government are aimed at supporting growth and integration with the global economy. The reforms have reduced licensing requirements and made foreign technology accessible. The reforms have also removed restrictions on investment and made the process of investment easier" (NASSCOM 2002).

15. For a detailed discussion of the values of Infosys, see http://www.infosys.com/brochure/Values_New_v2.pdf (accessed April 17, 2007).

16. Available at http://www.time.com/time/business/article/0,8599,1943185,00.html (accessed October 29, 2012).

are not allowed to get away with it; quick corrective actions are taken and I think . . . authorities are doing that."[17] Subsequently, he also defended India's standing in the global capital network by arguing, "One or two people have drawn attention to the fact that Satyam has been seen as a very negative development, but they appreciated that it is not only in India that these things happen, you have frauds in Europe, you have frauds in the U.S."[18] The latter assertion by Ahluwalia that the Satyam Scandal is not an isolated occurrence but is common to all nations is visible in how the debacle was also termed the "Enron of India."

Despite the anxiety stirred by the Satyam Scandal in terms of the outsourcing business environment in India, Murthy creatively used it to reinforce the ethical pledge of the industry, Infosys in particular. In an interview with CNBC-TV18, Narayan Murthy openly chided Satyam Computers and put to rest any speculations about Infosys buying the company. He unequivocally stated, "We will not touch such a tainted company."[19] In the wake of the scandal, Infosys also directed its HR to avoid hiring employees who had lost their jobs at Satyam.[20] The distancing from Satyam as a fellow IT company was unambiguous and complete. By singling out Satyam, Murthy also reinforced the ethical depth IT companies in India possess in general.

On the other hand, the case of the mining scam that consumed Karnataka till the end of 2011 offers us an interesting way in which the IT industry dissociates itself from the older industries as well. The mining scam eventually led to the ousting of the chief minister of Karnataka, B.S. Yeddyurappa, and his cabinet colleagues, including the Reddy brothers, B. Sriramulu, and V. Somanna. *The Hindu* reported the Karnataka Lokayukta (ombudsman) Santosh Hegde's indictment as follows:

> He [Santosh Hegde] said a mining company in Karnataka has donated Rs. 10 crore to a trust owned by the Chief Minister's family members. He added that the donation was made by the company for reasons other than genuine. The company also paid Rs. 20 crore for purchase of land by the trust far above the guidance value which makes for an offence under Prevention of Corruption Act.[21]

17. Available at http://articles.economictimes.indiatimes.com/2009-02-01/news/28478378_1 _satyam-scandal-satyam-issue-biggest-corporate-scam (accessed April 17, 2013).

18. Available at http://articles.economictimes.indiatimes.com/2009-02-01/news/28478378_1 _satyam-scandal-satyam-issue-biggest-corporate-scam (accessed April 17, 2013).

19. Available at http://www.moneycontrol.com/news/business/we-will-not-touch-tainted-co -like-satyam-narayana-murthy_375751.html (accessed October 29, 2012).

20. Available at http://articles.economictimes.indiatimes.com/2009-01-09/news/28483165_1 _satyam-employees-ceo-ram-mynampati-satyam-computer-services (accessed October 29, 2012).

21. Available at http://www.thehindu.com/news/national/lokayukta-submits-report-on-ille gal-mining-to-karnataka-govt/article2299141.ece (accessed October 29, 2012).

Reading ethnographically, the mining scam fortifies the IT narrative of ethics since it reveals the corruption of the state officials and the lack of transparency and accountability in the government apparatus. The mining industry, which belongs to the long-established industrial sector, in this context, is perceived as steeped in corrupt practices. These older sectors are perceived to be detached from the global market that necessitates the building of trust for capital to flow across national boundaries. It was quite common for employees at Infosys to think of the industrial sector as physically discomforting because of the heat, dust, and grime of factory work. They also saw the industrial sector as trapped in a time warp, while the world has moved on to more advanced modes of production. Such attitudes are especially interesting because most employees at Infosys were engineering rather than computer science graduates, whose formal education otherwise was geared toward industrial production. Further, the middle-class descent and the education of the newer entrepreneurs, such as Murthy and Nilekani, is seen as "naturally" leading them to embrace ethical courses. The Reddy brothers, on the other hand, came from more humble backgrounds and were working as activists for the Hindu fundamentalist parties, Bharatiya Janata Party, before they turned their fortune through illegal mining. The middle class collectively perceives the lack of education, lower-class-caste descent, and rural upbringing as a crucible for illegally amassing wealth.

The "Outside" of IT

It was August 2004, and I was in Bangalore for my extended fieldwork. My first appointment for the day was with Nandan Nilekani. The website of Infosys Technologies then described him as follows:

> [He is] the recipient of several awards including the Indian Institute of Technology (IIT), Bombay's "Distinguished Alumnus" award in 1999. In addition, Mr. Nilekani, along with Infosys Chairman Mr. N. R. Narayana Murthy, received the *Fortune* magazine's "Asia's Businessmen of the Year 2003" award. He was named among the "World's most respected business leaders" in 2002 and 2003, according to a global survey by *Financial Times* and Pricewaterhouse Coopers. Mr. Nilekani was also awarded the Corporate Citizen of the Year Award, at the Asia Business Leader Awards (2004) organized by CNBC. In 2005, Mr. Nilekani was awarded the prestigious Joseph Schumpeter prize for innovative services in field of economy, economic sciences and politics.[22]

22. Infosys Technologies Limited, Management Profiles: Nandan M. Nilekani, available at http://www.infosys.com/about/Nandan_Nilekani.asp (accessed November 21, 2006).

His biography is situated at the intersection of business, politics, and society that evidently go beyond the limits of the IT industry. I entered Building 1, which houses the corporate headquarters.[23] The receptionist sat at a desk in the center of the spacious marble lobby; the walls were decorated with original paintings. Once I introduced myself to the receptionist, she confirmed my appointment and asked me to take a seat. As I was shuffling through the magazines I found on the coffee table, I suddenly heard someone call out, "Lights, camera." I turned to the direction of the voice and saw a television crew at work. Nilekani was sitting on a chair, his face flooded with dazzling golden light, attentively answering his host's questions while looking into the camera. The receptionist informed me that the interview would air on a prime-time show that week. Nilekani would be heard talking about India's impressive IT global trade and the promise it ultimately holds for the nation.

After a while, I was directed to Nilekani's office on the second floor. I sat in the waiting lounge, rallying my thoughts. I was offered a glass of water, a practice common to welcoming a guest especially during the hot summer months in India. His secretary, Mallika, announced me. Nilekani's office carried numerous communication devices: with his secretary who sits outside his office, with other company personnel both in Bangalore and elsewhere in India, with satellite offices around the world, and with his present and prospective clients in the United States, Europe, and Australia. As we started speaking, he meticulously answered my questions regarding the people and technical processes of the IT industry, carefully defending the fact that IT is not a "routine" job as one would tend to think but involves creativity to "solve a customer's problems." However, his account of the IT industry slowly crawled beyond Infosys, and at some point he started using IT as an anchor to talk about India, specifically the disgrace produced by the "failure" of the state.

His concerns lay past the everyday routine of software development work: "corruption," social development, and India's niche in the era of globalization not only as an IT destination but also an emerging power. It is important to note how he persuasively talked about the several-layered realms, which to him are inseparably connected to IT as an industry: IT, in this context, is not declaimed by technology alone. To Nilekani being passionate about IT is being passionate about the promise IT holds for India through what he proudly calls its "multiplier effect": "'IT has the potential to modernize the society and make India a leader among nations." I gathered from his words that IT has a wider stake than fulfilling mere business expectations; it cascades into

23. All the buildings at Infosys are numbered for identification. They do not have any names as one would possibly expect, but the conference rooms in Building 1 are named after Indian Nobel laureates and other scholars and scientists across the world.

the domain of the nation-state by leveraging its internal logic of success. I asked Nilekani to elaborate on his understanding of the transformative nature of IT, particularly identifying the stakes involved: "So many things can go wrong. . . . India has a huge role in the knowledge economy to be global supplier of a wide range of services that has a potential to drive economic growth and this can go wrong if we do not give it adequate support, and become complacent and lose out in competition with other countries."

Nilekani is not alone in this. Narayan Murthy was also significantly invested in and advocated a similar vision of change that employs the success of IT as a catalyst. Similar to Nilekani, Murthy is also profiled in ways that indicate his active commitment to issues beyond IT:

> Mr. Murthy is the recipient of numerous awards and honors. *The Economist* ranked him eighth on the list of the 15 most admired global leaders (2005). He was ranked 28th among the world's most-respected business leaders by the *Financial Times* (2005). He topped the *Economic Times*'s Corporate Dossier list of India's most powerful CEOs for two consecutive years—2004 and 2005. . . . He was awarded the Max Schmidheiny Liberty 2001 prize (Switzerland), in recognition of his promotion of individual responsibility and liberty.[24]

In an article published in the *Times of India*, a leading national daily, one can note Murthy's presence in the wider public imagination:

> Infosys chairman, Narayana Murthy, is no ordinary man. After guiding Infosys through the boom and bust of the IT cycle, Murthy may have more than just the company on his mind. In an exclusive interview to the *Times of India* at the India economic summit 2001, Murthy said India needs proactive political leadership, enlightened bureaucracy, and a corporate leadership with a social conscience to solve many of the problems affecting the country. At the same time, he doesn't want to step in yet. This, despite his deep interest in alleviating poverty, bringing education to the masses and bridging the gap between the rich and the poor. "Sitting at where I sit, looking at what I look, I have no desire to be a politician. But god is omniscient. So, who knows," he replies to a query on whether he would like to become a politician.[25]

---

24. Infosys Technologies Limited, Management Profiles: N. R. Narayana Murthy, available at http://www.infosys.com/about/Nandan_Nilekani.asp (accessed December 19, 2006).

25. Times News Network, December 2, 2001, available at http://www1.timesofindia.india times.com/cms.dll/articleshow?art_id=734083526 (accessed December 24, 2006).

The media is replete with similar interviews of IT entrepreneurs who, while talking about their own companies, also comment on the state of affairs in the country and recommend modes of mitigation that one can and *should* pursue in public governance. Besides the media that caters to the wider public sphere, such rhetoric is also dominant in exclusive business domains. In an address at a gathering of more than eight hundred chief information officers from around the world attending the NASSCOM 2003 India Leadership Forum, Murthy offered the following in his speech: "I dream of an India when people overseas respect us because we are Indians. They will do this because we solve our problems quickly, [and we are] secular, global and modern."[26]

Referring to this speech a few months later, the same news service carried an editorial titled "Is It Time to Mentor the Nation Mr. Murthy?"

> The preceding quotes are not from any clear sighted statesman but from NR Narayana Murthy, chief mentor of Infy.[27] . . . Looks like Murthy is priming up for a role amongst nation-builders. Rumors abound that he is likely to step down from the Infy board and take up full time political assignment. . . . Is it time the chief mentor of Infosys Technologies turned mentor to the nation?[28]

When I first met him, Murthy evaluated the state bureaucrats as even "worse than the colonizers" because "they think of themselves as the elites and the government belongs to them." He mentioned Frantz Fanon, taking care to mention *Black Skin, White Masks*, "a book I read when I was living in France," to reinforce his argument. He further elaborated his opinion of Indian bureaucrats by arguing, "They are the owners and the managers and this colonial thing carries on even after independence. It is something about civil society that makes this enduring." He identified the specific nature of the problem in a comparative mode: "There are three things that one needs to follow in order to run a company: fairness, transparency and accountability . . . all of these are very low in the government." In the business world, these words are the basis of corporate governance, but they were intently embraced by Infosys to describe the "pillars" on which the company is founded. In addition, Murthy stressed that he undoubtedly sees a strong parallel between governing a corporation and governing a nation because "as much as the Infosys employees

---

26. Cyber News Service, February 11, 2003, available at http://www.ciol.com/content/opin ion/103021101.asp (accessed December 24, 2006).

27. Infosys is also referred to as "Infy" mostly as a short form, but sometimes also with fond ness.

28. CIOL Bureau, June 13, 2003.

or Infoscions[29] as they are referred to, are a stakeholder in this company so are the citizens of India, they are stakeholders of the country as well."

He emphasized the urgent need to reform the bureaucracy in order to ensure the progress of India while recounting the "difficulty" his company faced in the early 1990s when they had just started working on software development outsourced to them by clients in the United States. Most of this had to do with "red tape" and also with the "cavalier" nature of the administrators. He is "glad" that a lot has changed since then with new state policies that not only eased the flow of goods (software and hardware) and human resources in and out of the country but also evidently changed the attitude of government employees: "Now I can pick up the phone and get the job done."[30]

Toward the end of the conversation Murthy proceeded to add a word of caution: "But remember, we are not a good example. You should talk to people, outside." By "outside" he implied "outside Infosys" and, by extension, the "outside of the IT industry." The generic use of "people" en masse as opposed to identifiable Infosys employees he draws a troubling, but an intriguing boundary nonetheless to gauge the impact of the success of IT. Nilekani, as I mentioned above, was referring to a similar dichotomy of what is contained inside IT and what lies outside, but a dichotomy that is structurally connected through what he calls the "multiplier effect." The inside-outside dichotomy is central to my analysis here for two reasons: first, the *outside* is of more significance compared with the inside because to the likes of Nilekani and Murthy, the question here is of a wider social transformation rather than an isolated instance. Second, the dichotomy becomes problematic once we start analyzing the contours of its implications for the wider social domain. The outside, contrary to the aspiration of the IT entrepreneurs, is not homogeneous and neither is the rhetoric of change oblivious to existing social hierarchies.

---

29. I refer to the employees at Infosys as "Infoscions," which is a nomenclature they preferred over the generic software developers or engineers. Rather than taking away from a critique, this term helped me convey the parameters through which software engineers view their own work and its relation to the nation-state. Infoscions carry significant social capital "outside" Infosys, that is, in the city of Bangalore, where younger employees were often seen, especially in the emerging lifestyle malls, sporting their employee IDs around their necks as they were required to wear while at work. As some employees mentioned to me it was either to attract the opposite sex or to demand special treatment from the retailers. In any case, they wore it with pride, and most of the time "it worked" to garner favors. Even the ID card Infosys issued identifying me as a "Researcher" proved pragmatically valuable. I had to provide my landlord a letter from Infosys and attach it with a copy of the ID card before he agreed to rent me his apartment. When I had initially contacted him from the United States, he had mentioned that he "only rents to professionals who work in IT because then I do not have to worry about people defaulting on their rent." Also, I was able to secure a landline phone connection from a private operator within hours of providing my ID card.
30. Murthy cites the example of a Chinese student who was recruited for an internship at Infosys and was being denied a visa by the government of India; it took Murthy to make only "a phone call [to the appropriate state department] and the visa was granted in the next two days."

Rather, the way in which change is designed, for example, in the Water Project it bellies the monolithic outside and also the all-inclusive ethic of the rhetoric.

The Network

The network between Infosys and Janaagraha was reasonably well established by the time I was doing my fieldwork. While Infosys had developed civil society partners, mainly in the form of middle-class NGOs, I consider the alliance between Infosys and Janaagraha as paradigmatic since it captures the ethos of the market quite seamlessly across the two seemingly distinct areas—IT and civil society. One will recall that before launching Janaagraha, Ramanathan had worked with Nilekani, then chair of the BATF. Armed with the slogan "Bangalore Forward," BATF was envisaged as

> a unique experiment in private-public partnership in urban governance, where a new model of engagement has been tried, and has been successful in demonstrating the need for more private public partnerships for urban development issues. . . . In a scenario where liberalisation [*sic*] is a catch word, and the private and public sectors are increasingly working closer together for economic growth, development and improvement of the quality of life of people, the BATF was an idea whose time had come.[31]

The raison d'être of BATF was as follows:

> The rapid urbanization [*sic*], population growth and growth of the urban economy and industry of Bangalore has placed an enormous strain on the city's infrastructure. This led to the usual problems of a developing metropolis such as garbage, traffic congestion, deteriorating roads, pollution, and strain on civic supplies like water, drainage and electricity.[32]

The primary focus of the BATF was on the city's infrastructure. The city, the BATF members argued, was reeling under the increasing pressure of population growth owing to the opening of new industrial sectors, IT being the primary employer. As required by the public-private partnership mandate, the members of the BATF were an amalgam of government officers and heads of corporations, mainly from the IT sector "committed to civic-reform." It

---

31. "About BATF," available at http://www.bangalorebio.com/externalsite.htm?http://www .blrforward.com (accessed March 13, 2002).

32. "BATF: The Genesis—The Promise of Change and Action," available at http://www.banga lorebio.com/externalsite.htm?http://www.blrforward (accessed March 13, 2002). Emphasis mine.

received considerable amount of donations from these corporations as well as from individual members. For example, Infosys Foundation itself donated Rs. 63.55 lakhs (approximately US$101,000) for 1,115 push carts and six auto tippers for the "Swachha (Clean and Clear) Bangalore" Phase II run by the BATF; Narayan Murthy and his wife, Sudha Murthy, donated Rs. 8 crores for "Nirmala (Clean) Bangalore Sanitation System"; and Nandan Nilekani and his wife, Rohini Nilekani, constituted the *Adhar Trust*[33] to support Bangalore's development.[34] The Software and Technologies Parks of India (STPI) adopted fifty parks in the city for maintenance and development for a period of three years.[35]

It was also during the BATF regime that Singapore, considered to be a model world-class city, became an important reference for urban renewal in Bangalore (Roy and Ong 2012). The aspiration was popularized in catch-phrases such as "Let's make Bangalore into Singapore." Consultants from Singapore were contracted by the Bangalore Development Authority (BDA) to review the Comprehensive Development Plan (CDP) of the city. Jurong, the consulting firm, proposed the "IT Corridor" that stretched from the Electronics City in the south to International Technology Park Limited (ITPL) near the Old Madras road in the northeastern part of the city. This stretch of about 25 kilometers (15.5 miles) in length and 7.5 kilometers (4.65 miles) in width and covering an area of about 138.6 square kilometers (83.16 square miles) where most IT companies would be located was designed to "showcase (an) environment for IT professionals to live, work, play and strike business deals."[36] As Subroto Bagchi, chief operating officer of MindTree Consulting, stated: "At the end of the day, running cities like New York or Singapore and Shanghai is specialized [*sic*], full-time work. Bangalore cannot be an exception."[37]

Janaki Nair (2000) has written extensively about the Singapore-Bangalore dyad, particularly highlighting the misplaced nature of the comparison. She identifies several ways Singapore cannot be replicated in Bangalore for geographical, historical, social, and cultural differences. However, one of the important issues she highlights in this respect is the difference in the commitment between the two city administrations to public services such as housing.

---

33. The amount donated to this trust has remained undisclosed to the public.

34. Available at http://www.bangalorebio.com/externalsite.htm?http://www.blrforward.com (accessed August 1, 2006).

35. Available at http://www.bangalorebio.com/externalsite.htm?http://www.blrforward.com (accessed August 1, 2006). A new industry that was involving itself in the BATF is the biotechnology industry, which is professed to be the next major industry after IT. Biocon, the best-known biotech corporation in Bangalore, made a contribution toward the BATF initiative, the amount of which was not disclosed.

36. CDP, Bangalore Development Authority, 2003.

37. "Is the Bangalore Story Over?" *Business Standard in New Delhi*, September 21, 2005, available at http://in.rediff.com/money/2005/sep/21spec.htm. Emphasis mine.

In her aptly titled article "Singapore Is Not Bangalore's Destiny," Nair argues that despite the shared colonial pasts between the city and the city-state, "if Singapore today is acknowledged as no mean achievement, it is because it fashioned a future adequate to its own historical and geographical contingencies" (2000: 1513). For instance, Nair identifies public housing and mass transit as two of the contingencies attended by the Singapore government in rebuilding the city. These two initiatives were not only absent in Bangalore; rather, the trend was toward private ownership of residence and automobiles, which are markers of middle-class consumption practices. On the other hand, whether Bangalore ought to follow the steps of an undemocratic regime as demonstrated by Singapore is another important issue that one needs to examine in the light of the history of the city.[38]

Nair also points to the urban-rural continuity of Bangalore, which is absent in the case of Singapore. When we consider that the IT parks were previously agricultural land that was consumed by the entry of global capital, the nonexistence of the rural in the urban discourse of Bangalore is rather disturbing. Infosys has provided jobs to those who were displaced from their land and agricultural livelihood, employing them as the tour guides/gardeners I met. One tour guide in particular drove me around the campus with a sense of pride in being part of what he termed "something big." He could not articulate what this "big" entailed, but it was a code for him to think of Infosys as a space where "foreigners come and important things get done." Considering that he had no access to any of the buildings, his participation was peripheral and limited. But in the capitalist narrative, this participation is seen as benevolence for the global stage to be maintained. In her exploration of the development of Bangalore as a megacity, Gayatri Chakravorty Spivak argues that "rural is the interdiction of the local and the global-in-urban-space . . . the megacity initiative in all its forms, tries to lift the interdiction, gives the urban a 'proper' access to globality via the electronic, and transform the 'rural' into a metaconstitutive outside for the 'urban'" (2000: 18). In Chapter 2 we will see the transformation of the rural or what had come to be known as the peri-urban, where I show how the interstitial/peripheral territory becomes a space of neoliberal experiments.

By the time I was conducting my yearlong fieldwork in Bangalore in 2004–2005, BATF was a defunct organization. Dharam Singh, then chief minister of Karnataka, who came to power in the 2004 elections, discontinued

---

38. In 1990 Lee Kwan Yew, the first prime minister of the Republic of Singapore, took the title of "mentor." Murthy adopted the same in 2002. For both men the title indicates a sense of achievement as the keeper and preceptor of the city-state and the corporation, respectively, at a time when these two sites are inseparably connected in a neoliberal world. Having a mentor, especially in a world-class city, is about locational capital; it signals success at the global level, which is beyond the muddle of national and/or regional politics and the mundane engagements of the state. Between August 2011 and May 2013, Narayan Murthy was the chairman emeritus of Infosys, which has obvious academic resonance. He is currently listed as the "Founder."

it. The discontinuation of the BATF proved to be a significant worry among the IT entrepreneurs who saw it as a threat to their ambition to display the city as reliably equipped for business to clients from the United States and Europe. Drawing on the experiences of his prior involvement with BATF and inspired by his own "passion to change urban India," Ramanathan founded his own organization, Janaagraha, in 2002. Janaagraha was envisioned as a departure from typical civil society organizations such as NGOs; its unique nature was described as "a citizen's movement based in Bangalore committed to increasing citizen participation in local government: the practice of partici-patory democracy. It runs . . . campaigns that engage citizens, government, non-governmental organizations (NGOs), and corporate institutions, each stakeholders with an interest in a better Bangalore."[39]

The Ramanathans often explained that Gandhi's idea of "Satyagraha," or the "force of truth,"[40] which he embraced during the nationalist struggle, inspired the name. "Janaagraha means the Life Force of the People: it stands for a positive, constructive firmness to allow citizens to engage with their gov-ernment on specific issues. To provide a platform that creates the day-to-day successes that we need. To remove the cynicism from our minds. To re-instill hope in our ability to build a great country."[41]

The anxiety generated by the termination of the BATF found some relief in the programs launched by Janaagraha, such as "Ward Works," "Ward Vision Campaign," and "Taxation Campaign." However, Janaagraha's work went beyond infrastructural management to enter domains of urban fiscal governance, as well. For instance, along with some other NGOs (one of them called *Akshara* [alphabet], headed by Rohini Nilekani), Janaagraha initiated a new program called the Public Record of Operations and Finance (PROOF), introduced as follows:

> Performance audits and quarterly financial statements are universally acknowledged as essential mechanisms and a criteria for progress. The corporate sector, the NGO world, CBOs and civil society have not only embraced the concept, but have used it as the basis of perfor-mance measurement and the springboard of good governance. Today, we need the Government to practice it, PROOF provides this plat-form. It is about our Government building confidence with PROOF.[42]

---

39. "What Is Janaagraha?" available at http://www.janaagraha.org/about/about_us.htm# develop (accessed October 8, 2004).

40. In drawing upon Gandhi's Satyagraha, Janaagraha as a concept not only appropriates the memory of the historical struggle but, in the process, also displaces the meaning of struggle itself.

41. Available at http://www.janaagraha.org/about/about_us.htm#develop (accessed Octo-ber 5, 2006). Emphasis mine.

42. "What Is PROOF?" available at http://www.janaagraha.org/campaigns/proof_paper.htm (accessed October 8, 2006).

Nilekani personally endorsed PROOF and took an active role in the public event where the Bangalore Mahanagar Palike (BMP),[43] the city administration, announced its financial performance every quarter. He declared, "I support the PROOF campaign wholeheartedly. As a CEO I cannot imagine running my organization without credible information being produced, disseminated and used on a regular basis by all stakeholders. . . . Especially in today's climate such information is more than just performance, it is about fundamental institutional integrity."[44]

While the rhetoric and practices of the BATF and then of Janaagraha are focused on the city of Bangalore, in effect they go beyond the city. The reform is *of* the city but is *for* an external audit primarily by a potential business client from the West. The issue was not merely about aesthetics or efficiency; it also conveyed a sense of ethical integrity. The renewal of Bangalore is critical mainly to purge its seeming paradox: a globally successful city despite its location in a developing country. With the rise of the IT industry and the influx of IT workers from all over India, the city is currently burdened with myriad infrastructural problems. Its rapid transformation from the quiet "Pensioner's Paradise" to the neoliberal capital of the nation has impacted both the topographical and the ideological terrain of Bangalore. BATF, and later Janaagraha, focused on the infrastructural inadequacies—street conditions, water supply, garbage collection, streetlights, and the like—as entry points for talking about reforming public governance and overseeing the use of public funds by state administrators.

From several conversations at both Janaagraha and Infosys, I distilled some identifying elements of the dissatisfaction with the city infrastructure: it started with how "awful" the commute to work is given the potholes and the rising traffic congestion one has to negotiate. Next were the narrow streets and the increasing number of privately owned vehicles and finally the older industrial sectors (such as the silk and textile industries) that took up vital city spaces pushing the more important ones like IT to the peripheries of the city. The urgency to address infrastructural issues was, however, not confined to simply easing the daily commute of the city dwellers alone. It was also to, as one informant at Infosys put it, "make it bearable to the client from the U.S. who just landed"—the airport road was apparently notorious for traffic congestion throughout the day. Most of my informants at Janaagraha would say, "Something has to change. Bangalore has to look different." There was definite emphasis on aesthetics, efficiency, and ethics to portray Bangalore to the foreign client as ready for global capital.

---

43. In the PROOF Public Discussion Quarter II-2004-05, December 11, 2004, in Bangalore that I attended, Nilekani was the moderator for the question-and-answer session between the BMP officials and the citizens.
44. "PROOF Booklet, 2002," available at http://www.janaagraha.org/campaigns/proof_paper .htm (accessed October 8, 2006).

Performing Bangalore

The simulation of Bangalore as the "Silicon Valley of India" is rather convoluted: while emulating California, the moniker also aspires to retain India not only in its name but also as a way to reclaim the lost pride of nation. The coupling of an idea with a geography is not necessarily incongruous. I want to make three specific claims in this context: First, performance entails a detailed everyday preparation and perhaps a finale. Related to the above, it is more about the expectation of business clients arriving in the city from the West rather than the actual. It is the expectation of the audit that lends urgency to prepare the city along the lines of what is deemed to be "world-class." Second, performance is a way to banish elements that may interfere with the world-class city script. It is also performative because the narrative regarding the failing civic amenities of the city involved a strong visual renovation. It gravitated around architecturally refurbishing the landscape of Bangalore with glass-walled office buildings, malls, cafés, and American fast-food chains such as McDonald's, KFC, Subway, and so on. The performative aspect was important at Infosys, as well. Every visitor is encouraged to take the tour, and the tour is often integrated into the itinerary of the day when clients visit. My tour involved stops at the "world-class" facilities the company has in order to ensure quality work, and also at the gym and the swimming pool meant to promote the well-being of employees. My tour guide also pointed out the "multicuisine" cafeterias, driving home the international gastronomical leanings of Infosys employees, such as the "American salad bar."

On the other hand, the sight of derelict slum and footpath makeshift settlements were seen as an impediment because they relentlessly reinforced the "third world" status of India to the world. However, this is rather paradoxical: Bangalore, being a city undergoing a significant real estate boom, draws construction workers from all over the country, who often establish their makeshift dwellings on the sidewalks, usually close to a public water fountain. The shanty huts, as volunteers in Janaagraha often argued, ruin the world-class image of the city, yet this informal labor force is indispensable to this ambition.

So the question arises: What is a "world-class" city beyond the presentation of sanitized spaces? More importantly perhaps, what kinds of exclusion does this term necessitate? Would it not suffice for Infosys to showcase their manicured campus to their clients and not agonize over sanitizing the city? Would the guided tour and the power lunches not convince the client that they could trust Infosys with the project they were planning to outsource despite the potholed roads they traveled? Why and how do Murthy and Nilekani make the "outside" germane for the "inside" of their world of business deals? Why does the city matter even when most of these companies are located beyond the city limits in spaces dedicated exclusively for the IT (and biotechnology) industries? I want to interrogate not just how Infosys connects with the outside but what is at *stake* in this connection.

Thomas Friedman, in his *The World Is Flat*, recounts his interaction with Nilekani, who tells him at one point (when the TV crew was setting up the cameras in Nilekani's office for the Discovery Channel series on globalization): "Tom, the playing field is being leveled" (2005: 7).[45] Friedman goes on to interpret this as: "He meant that countries like India are now being able to compete for global knowledge work as never before—and that America had better get ready for this." From Nilekani's phrase, "the playing field is being leveled," Friedman constructs his brisk thesis of the world that turned "flat" "while he was sleeping." It also points to the kind of dramatic imagination of the city that the likes of Nilekani and Murthy have produced for their revered Western audience. Friedman's visit as a journalist from the United States to Bangalore, and his eventual volume, was a corroboration of the desire of the IT entrepreneurs to showcase Bangalore as a global city. Copies of the book sold with unusual rapidity as soon as it hit the bookstores in Bangalore. It topped the best seller list published by the Indian dailies for more than a month.

Mr. Shanbad, owner of Premier, an independent bookstore, and someone thoroughly uninfluenced by the neoliberal prospect of Bangalore,[46] was also taken by surprise at the record sale of this particular volume. As a longtime committed bookseller in a "city that seldom reads anything other than codes," he interpreted this sale as indicative of a "ready set of people who were eager for this."[47] He and I compared the sale of *The World Is Flat* with historian Janaki Nair's (2005) book, *The Promise of the Metropolis: Bangalore's Twentieth Century*, an in-depth analysis of the city, which though published around the same time did not find the same audience in the city. The enthusiasm of the readers was above all to find a validation of the "new economy," IT's crucial role in it, and their city as its center, by none other than a Western writer. I do not think it is a stretch to argue that *The World Is Flat* was a defining moment for the emerging neoliberalism in India where the West finally took note of this momentous transformation.

In the days following the release of the book, almost every Infosys employee I knew purchased a copy. Most were deeply pleased with Friedman's treatment of the theme of a "changing India." As many of them put it, "We are proud to be driving this change as IT people." The *New York Times* has been publishing stories about foreign cars, real estate, and highways that index an

---

45. Emphasis mine.

46. Most bookstores in the city were undergoing "look and feel" renovation like Barnes and Noble and Borders in the United States, with cafés being regularly added to complete the experience. Premier, on the other hand, although located in downtown Bangalore and mostly carrying serious works, continued as a dusty store with shelves piled up with books that precariously and strangely sustained themselves without crashing to the floor.

47. He refers to software codes.

emergent lifestyle narrated mostly in terms of economic achievement. One such case is Amy Waldman's series of articles carried by the daily in December 2005 titled *India Acceleration*.[48] While Waldman does refer to the "road more traveled," most people in India are still forced to ride in overcrowded buses, the temper of the articles definitely carried a changing, more affluent India that is gradually emerging.

In this discourse of a transforming nation, Bangalore *is* the epicenter from where the playing field of the globe is being leveled. It is the center from which a new India is to emerge. It is interesting to note that state visits of foreign diplomats started including an obligatory visit to Bangalore, sometimes to the extent of making it the first stop, even before New Delhi, the country's capital, as in the case of the Chinese prime minister Wen Jiabao in April 2005. During his visit, Wen Jiabao, in an attempt to foster business links between India and China, said, "Co-operation is just like two pagodas. One, hardware and one, software. Combined we can take the leadership position in the world."[49] To understand this simply in economic terms tells only half the story; it carries deeper implications for a new way of thinking of the nation-state and the frantic attempt to display Bangalore as the face of an emergent India that is not only open to the global market but also to the global spectators, all of which is to ensure the country a prominent place in the geopolitics.

Recently, the aspiration of Bangalore to be a place for global capital and individual freedom to flourish has been somewhat compromised. The city has experienced incidents of harassment of and crimes against women, especially those engaged in call centers. Call centers mainly serve the clientele of companies based in Europe and the United States, who are located six to twelve hours of time difference from India. This implies that the usual working hours for the outsourcing industry in India are overnight to intersect with the daytime hours in the Western Hemisphere. The overnight schedule, among other things, such as disarraying the circadian rhythm of the workers, has led to crimes against women as they commute to and from work. It is important to mention here that these incidents have occurred despite the effort made by most companies to provide transportation, especially to the female employees as they are venturing out at an hour that is culturally deemed unsafe and unsavory for women. The debate around the safety of women in the outsourcing industry has brought the question of public space out into the open in relation to women's professional ambitions. It seems to me that these crimes indicate the continuity of the public space in India as a male space, only selectively available to women during certain times of the day. Also, considering that IT or the call centers are predominantly male

---

48. Available at nytimes.com/asia (accessed July 31, 2006).
49. Available at http://www.friendsoftibet.org/mediaonfot/2005.04.12-china_india_can _lead_world_in_it-afp.html (accessed July 31, 2006).

domains, such crimes could be seen as an extension of the patriarchal values, which now impact women in a new way.

Often in the public discourse, the women who work in the outsourcing industry, especially in the call centers, are considered sexually promiscuous since most of them live independent lives and liberally engage with men prior to their marriages. Bollywood films have captured this changing terrain of gender relationships, but the subtext continues to disparage the woman as either the architect or the victim of the harassment. Now whether this impacts the IT sector, I have serious doubts, primarily because gender was never a point of discussion in an otherwise male domain such as Infosys. Rather, Infosys often promoted the gendered roles—man as the prime breadwinner—in hiring and retaining employees. As a corporate ethos, nevertheless, Infosys perceived itself as a gender-neutral space since I found it almost impossible to discuss gender as an issue in the workplace.

The other event that shook the image of the city in August 2012 was the mass exodus of workers originally from the northeastern states, especially Assam.[50] In August 2012 a text message stating that northeastern workers and students in the city would be killed in retaliation of a recent Bodo-Muslim conflict in the eastern state of Assam led workers from the northeast to flee. Though the state attempted to communicate that the text was most likely a rumor and guaranteed safety in return, the workers were not dissuaded. When one considers that the population in the northeast part of the country was never really an integral part of the country, their fright reveals the fracture of the nation-state. Though the connection is not direct, here it seems useful to think of the linguistic strife between Kannadigas and Tamils that has framed Bangalore for several decades now.

Drawing on Janaki Nair, the discord between what is generally termed "Kannada nationalism" and "Tamil nationalism" in the city relates to the migration of Tamils from the neighboring state of Tamil Nadu to work in the cantonment during British rule and the eventual replacement of Kannada by English in the official discourse. Tamil continued to have a strong cultural presence in the home state (Nair 2005). Nair argues that the displacement of Kannada is significantly related to the economic opportunities, which are increasingly being decided by the current global market that privilege English. In bringing this linguistic anxiety into our discussion, I want to suggest that the exodus of the workers from the northeast, despite state claims that the text message was possibly a sham, is plausibly embedded in the drawing of urban boundaries based on language. However, the linguistic strife in the city never surfaced openly during my work in Infosys or Janaagraha. These spaces

50. Available at http://articles.timesofindia.indiatimes.com/2012-08-16/bangalore/33232302_1_guwahati-rumours-top-cop (accessed March 12, 2013).

were perceived to be ascription-neutral, though it was not difficult to see how Infosys employees mostly gravitated toward people from their own states.

The Matter of IT

One question that has intrigued contemporary researchers is how and why IT as an industry has socially impacted India in unprecedented ways. Put differently: "Why and how does IT matter?" A considerable number of contemporary researchers in Bangalore studied what is usually known as the "social impact of IT," where the focus is primarily on computer literacy and the way people's lives are undergoing change in terms of how they are "empowered" through this kind of digital knowledge, particularly the Internet, to which they have access now. In these research studies, computer literacy is often seen as an index of "social development" or "progress." Scholars also focus on explaining how IT is representative of "globalization" and India's integration with the world economy. Still others delineate the transaction between the state and IT and the various policies framed since liberalization (Parthasarathy 2004). Finally, sociologists who are studying the IT industry from locations similar to mine, at places like Infosys, are documenting the "processes" at work, such as the "induction program," where the Infosys employees are introduced to the values of Infosys, or the various workshops held to train employees to adapt professionally and culturally to client sites in the United States and elsewhere. Their analyses mainly dwell on how a distinct corporate culture is being created in and through the IT industry, given the new demands of the global economy.[51]

While the software processes provide an excellent entry point to understanding the IT industry, limiting one's analysis to the processes alone obscures how the industry influences the "outside." Moreover, these processes are understood as "given" in the sociological analysis whereas the fact that they condense and conceal profound historical dynamics of notions about the "individual," the "social," and the "political" remains unproblematized. My extended discussions with Nilekani and Murthy, and the media reports about them, seemed to disclose something beyond just the "social impact of IT" or the "new corporate culture." They raised issues of a more entrenched nature. I contend that in a modernist sense, these processes are designed to erase ambivalence and ambiguity. One instance of this is the transformation of a new recruit to an "Infoscion" at the "induction program." The induction program is a mandatory orientation session meant to familiarize college re-

---

51. Most of the papers presented at the "International Conference on New Global Workforces and Virtual Workplaces: Connections, Culture and Control," August 12–13, 2005, situated IT within this perspective of a corporate culture.

cruits with the unique values of Infosys. The values,[52] collectively known as C-LIFE, are presented by one of the board members of Infosys:

- *Customer delight:* A commitment to surpassing our customer expectations.
- *Leadership by example:* A commitment to set standards in our business and transactions and be an exemplar for the industry and our own teams.
- *Integrity and transparency:* A commitment to be ethical, sincere and open in our dealings.
- *Fairness:* A commitment to be objective and transaction-oriented, thereby earning trust and respect.
- *Pursuit of excellence:* A commitment to strive relentlessly, to constantly improve ourselves, our teams, our services and products so as to become the best.[53]

The values, as one can see, have a generic quality. Yet they are applied to concoct a very unique setting, that is, to carve an inimitable "Infoscion." Rather similar to Ian Hacking's concept of "making up people," this is an attempt to *fix* the category of the Infoscion and then retrofit a corresponding set of people to occupy that category. At the end of these induction programs, I always asked the convener, "How can you ever make sure that they will imbibe and live these values?" More often than not, the reply was tentative: "Yes, there is no way to know, but we still have to try." I gradually appreciated the importance of investing in C-LIFE. It emerged from a modernist way of thinking where every human action and thought can be arranged in a legible way, to eliminate the possibility of ambivalence. It was also an important way to exhibit the ethical commitment of the company. The values, as most of the conveners agreed, are only tentative. However, they create an illusion of assurance that one has an available ethical repertoire to access should any doubt arise, both in professional and personal life. These values, as we can see, are rather generic, hence also *portable* between the inside and outside of IT.

The *portability* of these values back and forth between Infosys and the outside conjures IT as a *critique of the social*, as an ethical crucible from which a prescription and model for a wider social compact can arise. I employ "social" in a broader sense. It encompasses not only the transaction between the state and the civil society but also transactions within civil society itself. The civil society in this discourse, as Murthy pointed out, is as much a

---

52. This set of values was created at a two-day workshop, where, as a friend mentioned, consensus was reached after intense debate among the board members, senior officials, and some employees who were invited to participate.

53. Available at http://www.infosys.com/about/vision_and_mission.asp (accessed October 25, 2006).

"stakeholder" of India as the state is, but to understand the "stakes" involved, it is important that citizens recognize their renewed responsibilities. Ramesh Ramanathan constantly reminded citizens attending the Janaagraha meetings that the practice of "elect and forget" had to change to "elect and engage" political representatives.

To this effect, the rhetoric of change cannot be located as a critique simply of the Indian state—it is also aimed at citizenship. It attempts to enhance the notion of citizenship beyond the passive rhetoric of birthright to an active engagement with the state. The basis of this engagement is the new market economy where one's claim to the state as a citizen is legitimated as a valid "stakeholder" of the nation. Though used regularly at Infosys and Janaagraha, the concept of "stakeholder" continued to remain ethnographically blurred, particularly in the latter site. One of the questions I pursued was how does one understand "stake," an economic parameter, within the realms of citizenship, which carries significant affective dimensions? As I show through my analysis of the Water Project, the discussion of stake surreptitiously entered an ever-growing nebula especially when it shifted from the middle class to the urban poor.

Monday Morning Meetings

Tom Friedman interviewed Ramesh Ramanathan for his 2004 documentary *The Other Side of Outsourcing*, which preceded his book. Ramanathan spoke about the "urgency" to reform public governance, pointing to the fact that a few yards away from his office the slums were in pitiful condition and that they too were part of a city that is otherwise so inclined to display itself as the center of globalization. A basic Internet search revealed that Ramanathan had returned from the United States, "sacrificing" his job as an investment banker. Here it is interesting to note that Ramanathan deliberately talked about slums in Bangalore as his mission for urban renewal, when the dominant critique toward BATF, and later toward Janaagraha, has been that it was primarily intended for middle-class well-being. Over time I realized that evoking poverty and attending to issues of the poor, as I argued above, was essentially a collateral for the middle-class narrative to masquerade as classless.

To introduce myself at Janaagraha, I met the volunteer coordinator of Janaagraha, Sapna. We had already exchanged a few emails, and she was aware of my research project. Janaagraha was housed in a rented private residence in the northeastern part of the city with a signpost outside the iron gates that both my auto-rickshaw driver and I missed the first time. Colonel (Retired) Rudra, the office manager, welcomed me as Sapna called in to say that she was on her way. In the meantime, Colonel Rudra gave me some documents Janaagraha had produced on "citizen participation" and "reforming urban governance." Sapna arrived in the middle of my conversation with Colonel

Rudra and introduced herself while also apologizing for her delay, "having to negotiate the harrowing traffic every day."

Over a cup of sweetened tea in a separate room she began as follows:

> I will tell you what Janaagraha is and does and then put you on to others. We launched in 2002. It is a citizen's platform to engage with the government effectively. It is not a registered NGO. Meaning "life force of the people." Provides the tools and the processes to engage with the government constructively. Basically to improve governance on a community-level participation. Constructive engagement with the government. Starting with Bangalore and then extending it to the rest of the country.

She talked about the various programs at Janaagraha, such as the "Ward Vision Campaign," "Ward Works," "Taxation with Transparency," "Swarna Jayanti Sarkari Rozgaar Yojaney" (SJSRY, a state program to help the urban poor with their finances), PROOF, and the Water Project. She advised that I return next week and start by attending the "Monday Morning Meetings" at nine, which is a weekly report on all the programs and open to all. As Ramanathan had mentioned in our first meeting, Sapna also considered that the discussions in this meeting would help me identify the specific project I wanted to study and also provide an opportunity to meet the corresponding teams.

The meeting was attended by a motley group: full-time volunteers at Janaagraha, known as *Janaagrahis*, college students working as interns, young professionals, and also a substantial number of retirees. A clock was drawn on the whiteboard toward which everybody was facing, interspersed with acronyms at regular intervals, such as TWT, which stood for "Taxation with Transparency," and GBWASP or the Water Project. The meeting started with everybody, even those working at Janaagraha, introducing themselves, including Ramesh Ramanathan and his wife, Swathi Ramanathan. When my turn arrived, Sapna "took the opportunity" to introduce me as a "Ph.D. student from New York City interested in Bangalore and of course Janaagraha," looking specifically at the direction where the Ramanathans were seated. A mix of grins and knitted eyebrows appeared on the faces across the room as they chimed in to say welcome.

Each Monday one volunteer was informally selected to preside over the meeting. He or she started by introducing Janaagraha to those who were attending the meeting for the first time and then invited the project leaders to present an update of the activities from the past week and their plans for the present week. The sequence of the clock was closely followed unless somebody had to leave early for an appointment. While the leaders received questions from all, the ones that were encouraged were those from the longtime volunteers who were familiar with the various projects. The new attendees were generally encouraged to attend a briefing session that Sapna conducted shortly after the meeting, similar to the first meeting that I had with her.

I was inclined to choose between PROOF and the Water Project over the other programs, as I explain below. Once the meeting was over, and after two rounds of sweetened tea that day, I introduced myself personally to the leaders of these two programs, Mr. Srinath (PROOF) and Mr. Nayar (Water Project) and asked for their consent to attend their weekly meetings to acquire an idea of the projects. Mr. Nayar, usually referred to as KK, a retired technical manager from Hindustan Lever Limited, mentioned that the Water Project weekly meeting immediately followed the Monday morning meetings, and that I was "welcome to attend; we will start in fifteen minutes, at eleven." In the meantime, I sat in the middle of an array of activities. A volunteer was administering the Citizen Quotient (CQ) test[54] in one of the rooms, Colonel Rudra was advising his staff on the administrative jobs for the week, and there was a group surrounding Ramesh Ramanathan, waiting their turn to talk about problems they were facing with their projects, especially with state administrators.

Next week I went to Manjunathanagara, a slum in the then Greater Bangalore area, which would be part of the Water Project to familiarize myself with the site. It soon became apparent that though the residents had heard about the project, the details nonetheless eluded them, especially those involving the initial beneficiary contribution and the eventual charges that would be levied on water use. They were peripheral to the neoliberal prospects of Bangalore because of their everyday struggles around very basic necessities, water being one of them. Another issue, which is relevant for our discussion here and that was cited repeatedly by the slum residents, is the elimination of public transport. This harks back to the BATF decision to reduce public transport along certain routes to facilitate the commute of IT and other professionals. This severely affected the livelihood of many I spoke to in Manjunathanagara since they had no alternate means of transport to get to work. The middle-class reformation of citizenship and governance, though treated as a technical matter, is vulnerable to its own politics. This book is an attempt to represent the tangled nature of these two notions.

---

54. I took the CQ test that day as well, which I discuss in the next chapter.

2

Market Experiments
and Peripheral Lives

The CQ Test

On my first Monday at Janaagraha, a volunteer administered the CQ Test to me after the meeting. CQ stands for "Citizen Quotient."[1] With its clear derivation from the Intelligent Quotient (IQ), the test is known as the "Citizen Quotient Self-Test."[2] It was developed by Janaagraha and is directed at people who show interest in mending the state-citizen engagement. Since this is also a self-test, the tagline urges the test taker: "Be Honest. After all, it is a self-test. You got to keep the answers." The test, which is offered primarily in English, is divided into five main parts: A. Your identity as a voter; B. Your knowledge about your local government; C. Your interactions with the government; D. Your thoughts on how to improve the government; E. Your actions to improve the government. Each section was composed of subquestions that could be answered either yes or no. The questions in parts A and B were technical in nature, for instance, "Are you a registered voter?" "Have you ever voted for the local government elections?" "Do you know the name of your ward corporator (elected representative)?" The questions became increasingly ideological in the subsequent sections. For example, part C started with the query, "Have you ever bribed anyone in government

---

1. "CQ Self-Test," Janaagraha document, available at http://www.janaagraha.org/cq.htm (accessed November 21, 2006). See table 1.
2. Ibid.

to get your work done?" and ended with "Have you ever used 'contacts' inside government to get things done?" Since most test takers did pay a bribe or used personal connections, part D offered solution to the problem of "corruption," which ranged from indifference and fatalism to ethical responsibility, that is, "Get citizens to take responsibility." The last part is about possible actions that the test taker could consider, such as "have formed a network of citizens to work in partnership with the government" or "have taken several actions to fight corruption, not just for my problem."

The test taker personally tallies the responses. Only the yes answers are scored to highlight positive responses. The following scale was used to interpret the numerical scores as an index of citizenship:

> Less than 20—Your citizen quotient is at the bottom of the barrel. The good news is that it can only improve!
> 20–50—You still have a long way to go as a citizen.
> 50–100—You are a concerned citizen, but are you doing enough?
> 100–150—Well done! You are an active citizen.
> More than 150—Congratulations, you are a brilliant citizen.

The test was considered a rite of passage at Janaagraha. It was crucial for citizens to take this test to pass from a state of "oblivion" to a state of "awakening" about what citizenship actually entails, that is, an active knowledge of and engagement with one's government. As we can see above, the parts and the related questions, carefully disassemble civic life into comprehensible modules covering all possible democratic forms and practices of urban life in India. The test modules, as one of the Janaagraha volunteers explained to me, were designed to help citizens identify with the various ways they actually can and should engage with the government; "it's only that they do not realize it." Though the test was in theory said to be private, it took less than a couple of minutes before the test taker would divulge his or her score with quite a visible demonstration of guilt. Nobody during the length of my fieldwork scored beyond 50. The majority of test takers scored around 30, which, going by the above evaluation chart, is ranked as "poor" performance. The numerical value helped translate the citizen's location vis-à-vis the state and vis-à-vis fellow citizens: It made the *condition* of citizenship easily legible.

Though the test score was a surprise and an embarrassment to the test taker, Janaagraha more often than not expected the lower score and was prepared to work on the guilt that ensued. Usually a Janaagraha volunteer would respond to by saying, "So what do you want to do?" It would be followed by a dedicated session to introduce the various projects Janaagraha was involved in and the citizen could make his or her choice depending on skill set, educational background, and time he or she could spare every week. Given the ambience in Janaagraha and that the test was mainly available in English, it was not surprising that most of the test takers belonged to the middle class. They ranged from college students there for an internship period as part of their social service requirement, to young professionals who involved themselves in relation to CSR programs, typically from the IT companies, to retirees looking for some "meaningful" engagement.

A Shifting Cityscape

It is a sunny Monday morning in August and I am standing outside my apartment building trying to hire an auto-rickshaw to go to Janaagraha to attend the Monday Morning Meeting starting at nine.[3] After being graciously refused by eight drivers for reasons that they will not disclose, the ninth agrees to ferry me. The air gushing in through the open sides of the vehicle is nippy; quite a difference from eastern and northern India, where I had spent most of my life till then, where August is relatively hot. While the driver is comfortably negotiating the traffic, which seemed to be flowing in both predictable and unpredictable directions, my worries about making the meeting on time rise.

At the time of my fieldwork in 2004, Janaagraha was located on Nandidurg Road. Later in the summer of 2005, the office moved to the UNI building on Thimmaiah Road and was renamed Janaagraha Centre for Citizenship and Democracy. To reach Nandidurg Road, we started from Cox Town, where I lived, and passed through Frazer Town and Cooke Town, which were once part of the Bangalore Cantonment. The Cantonment established by the British about two centuries ago is populated mostly by people from the neighboring states of Tamil Nadu, Andhra Pradesh, and Kerala who are linguistically different from the western part of the city or *pete*, which is almost five centuries old and is primarily considered home to the Kannadigas, the linguistic group that comprises the state of Karnataka (Nair 2005). Kannadigas I personally knew often remarked that "the Cantonment is not for us," reinforcing the separation of the two linguistic areas in the city, which as I mentioned earlier has often been a source of conflict. However, the demographic profile of the area is changing quickly as IT is bringing in a relatively younger population from different parts of India, especially from the north and the east. The built landscape of the area is also shifting; the American-style supermarkets are replacing the traditional grocery stores while older houses with red-tiled slanting roofs are yielding to gated high-rise apartment buildings. Interestingly, as I was becoming familiar with the cityscape, I realized that these new buildings now provide crucial landmarks to navigate the city particularly when using public transport.

---

3. Auto-rickshaws are three-wheeled vehicles operating in urban and suburban India, which are often a less expensive alternative to hiring a cab.

That morning I faced a difficulty. As we entered Nandidurg Road through one of the back roads I was unfamiliar with, the auto-rickshaw driver wanted to know the name of the apartment building in the vicinity of Janaagraha, which would offer him a pointer. I did not have an answer. Though I had been to the Janaagraha office a few times previously, I had not realized that it was imperative to identify and know the names of high-rise buildings in a particular area so that they could be offered as landmarks when solicited. Just out of curiosity, I wanted to know what the past landmarks were. The auto-rickshaw driver, looking quite disinterested, said in his broken Hindi, "Nothing, madam, now it's a lot easier. Next time look for them. Every neighborhood has at least one, like the one where you live, Purva Park." By encouraging me to map the city in a certain way, he brought forth the altering setting of Bangalore and the new visual aid it offers in the form of new architectural interventions. However, his disinterest also revealed his social and economic distance from these expensive spaces. To him, like other auto-rickshaw drivers I met during my stay, these buildings had a topographical significance in terms of locating an address. The method worked seamlessly. From then on, when I had to return home, I would specifically say "Purva Park," the name of the apartment building where I lived, to auto-rickshaw drivers, rather than Cox Town or Banaswadi Road.

Janaagraha had a rule that Sapna mentioned to me in our first meeting: "Anybody who walks in even five minutes late to the Monday Morning Meeting has to pay a fine of five rupees." A plastic box would be religiously placed in the center of the table, and whoever walked in late would quietly deposit a five-rupee bill or a coin. It was meant to publicly and mildly shame tardiness, which is considered a national habit and popularized in Indian Standard Time (IST), often recast as "Indian Stretchable Time."[4] While pointing to efficiency, the practice was also intended to draw attention to the sluggish pace of the nation that needed to be altered. Every Monday the leaders of each project offered an update on their activities from the past week and their plans for the current week. These activities included meetings with state officials, organizing citizens' meetings in particular neighborhoods relevant to the project, planning for public meetings, and so on. I spent the first few weeks attending the weekly meetings of all the ongoing projects finally settling on the Water Project.

I start by arguing that if Janaagraha is indeed the civil society partner of the IT industry in reforming public governance and renewing citizenship, the presence or absence of the IT entrepreneurs is inconsequential. The currency of the market ought to be available to us ethnographically, independent of the actual presence of any member from the IT industry. The apparent absence of

---

4. Given that small change is scarce in urban India, sometimes latecomers and others exchanged higher bills for five-rupee notes.

the IT entrepreneurs in the Water Project is therefore central to my argument that links IT with its "outside." "Processual ethnography," as Paul Farmer (2006) has suggested, brings out these interconnections, which knit together otherwise disparate domains such as the IT industry, wider civil society, and NGOs. This method also offers us the opportunity to map the nuances of such emerging associations. I do not intend to conflate the two sites—Infosys and Janaagraha; rather I want to emphasize the common principles that animate their commitment to shape a new urban India. In both sites, a sense of urgency dominated the rhetoric to transform the socialist-redistributive model on which India is governed to one based on the market. At the same time, together they subscribe to the market as a tool to assuage this urgency.

The following are some of the questions I explore in this chapter: If the Water Project, as a neoliberal project, is designed to create a water market, how are concepts such as stake, transparency, and accountability positioned? How does the network between Janaagraha and Infosys become available to us ethnographically through the project? How does a comparison between two otherwise unrelated domains, water and software, help us understand the objective to reform public governance? Finally, how do we problematize the "outside" not just as a fractured reality but also how it turns our gaze around on IT itself?

Water Privatization and the Water Project

Water privatization has gained significant traction across the world, especially with the growing influence of multinationals such as Bechtel, Vivendi, Enron, and others. Coupled with these companies are the market reforms being endorsed by developmental agencies. My interest in privatization is in its emergence as process. This implies that an amenity and/or a service that had been traditionally delivered by the state at minimal or no cost would now be offered by a private provider with a clear motive for profit. This altering provision of basic amenity from a public to a private framework does not negate the state. Drawing on David Harvey (2007) here again, the state now enables and guarantees the realization of the private endeavor.

The literature on privatization, especially of water, usually takes a pro- or anti- position. Those who oppose privatization mainly do so on moral grounds, arguing that it is unethical to profit from the provision of a basic amenity such as water (McDonald and Ruiters 2005). The promoters of privatization on the other side usually argue that the private sector can provide water more efficiently than the state can (Franceys 2008). Central to both these arguments is, first, the question of providing water to the urban poor. Second, water is often positioned as a paradigmatic case for basic amenities in general. However, both these positions begin by looking at concepts such as "public," "private," and "civil society" as discrete and fixed categories, which in reality overlap. An analytical separation among them restricts the possibility of imagining an equitable future. Here it is important to take into account the dual nature of water, as a natural as well as a social entity both of which are embedded in invisible infrastructural networks. Further, water in the Indian context is also rooted in a religiously sacrosanct framework of life and death (Feldhaus 1995). To appreciate the complexity of water Karen Bakker invites us to "develop an understanding of the roles that both public and private actors play in governance of urban water supply for the poor, and pay close attention to the practices of urban water use in developing countries (particularly of those of the 'urban unconnected')" (2010: 6).

In this context, let us look at the keynote address at the inauguration of the Water Project in July 2005 by the then finance minister of the government of India, P. Chidambaram. He carefully coupled the role of the market and its ethics of accountability as a way to offset corruption:

> This is a unique event in which the concept of public-private partner-ship is being extended to municipal administration. In the years that we have talked about PPP, it can only be fully visible in an experiment like this. . . . The government loves to tax people. . . . Why? Because taxing people means raising money and not being accountable on a daily basis for that money . . . but raising money through bonds means to raise money that you are accountable for every year. . . . If you do not perform, the market will punish you. This I think is true accountability.[5]

The Water Project offers us a prism to explore the power play as a kind of imagination, a set of practices and a mode of inquiry into processes about governing water. Mainly, I study the project to interrogate the notion of the market at the center of this reform, its limits and possibilities, in its transactions among the state, Janaagraha, citizens, and international donor agencies. The Water Project has another critical aspect: its peripheral location that was at that time outside the jurisdictional area of the city of Bangalore. The Water Project covered the seven city municipal councils (CMCs) surround-ing the city. In early 2007 following protracted arguments, the CMCs were merged with the city of Bangalore. It was highlighted as the "mega-merger of Bangalore." However, the periphery, as I show, is critical for the experimental nature of the project.[6]

---

5. Personal recording.
6. Further, considering that the airport was also being built in the periphery of the city—Devanahalli—it is another instance of the politics of urban reform, which had a significant pres-ence of IT entrepreneurs. Here land is of prime importance, but a specific kind of land, that is, rural agricultural land. Gayatri Chakravorty Spivak (2000) has pointed to a very important connection between the rural and the global, real and the virtual. She argues, "For the fashioning of the general

I find Matthew Gandy's (2008) work on the peri-urban in Mumbai as a useful reference to think about its in-between location between the rural and the urban. In presenting us a picture of the outlying parts of Mumbai that are mostly unconnected to the main city water supply just like it was in Greater Bangalore, Gandy argues that such "disparities are etched into the urban landscapes: some of the slums with the worst service provisions are traversed by giant water pipes. . . . Unregulated construction activity and the provision of illegal water abstraction licenses have created intolerable conditions for poor communities" (2008: 116). Gandy invites us to analyze the disjuncture between the institutional framework of the postcolonial state as a set of rules and regulations and the departures from that framework that happen on the ground level frequented by forms of power that range from the public to the private and even personal. I extend Gandy's study of the peri-urban in arguing that its peripheral setting—adjacent yet separated from the city of Bangalore—lends a unique opportunity for neoliberal experimentation.

I use "experimental" to indicate, on one hand, the uncertainty of testing a new organizing tool: the market. On the other hand, I map the power relations that underscore such experimental zones. Together, these two aspects are constitutive of the neoliberal ethos. Theoretically, the market is a means to technically intervene and end corruption and institute an ethical state and engaged citizens. The assumption that underscores this instrumental approach is that the polity ought to transcend its familiar noisy political quagmire of corruption and distrust. Such an expectation is not only implausible to fulfill; it also leads to the emergence of a different kind of class politics that miserably isolates the poor as a "residual" and hence an undesirable socioeconomic group. The private operator will levy a charge for supplying water, but privatization itself is mainly about making a profit. The charge will not only be higher than usual. The profit collected will also no longer feed the existing redistributive state system that provided free water to the poor.

The Project

The Water Project is a rupees 650 crore (approximately $102 million) proposed water supply and sanitation project of the Water Board that will transport water drawn from the Cauvery River, almost 100 kilometers (62 miles) away, to

---

will for the real space of the megacity is in the interest of the virtual . . . by the incursion of foreign capital into the agricultural sector, making it even easier for foreign companies to buy land" (2000: 18). She discusses the "exploitative" land acquisition strategies practiced while planning the airport to raise the question of resistance. She refers to a hearing scheduled on January 13, 2000, to rationalize the need of a new airport in place of modernizing the existing one. Her informant, a labor activist and a law professor, had explained that the "small landowners had been completely won over by the megacity-prices offered by the Development Financial Institution. So the mere halting of the airport was no more than a skirmish. The bigger battle had been won" (ibid.).

supply the eight urban local bodies (ULBs) surrounding the city of Bangalore, known as Greater Bangalore. It was promoted as an innovation in financing of public infrastructure projects called the "Pooled Finance," where citizens were directly inducted as contributors. The details of the innovation are explained in *Pooled Finance Framework for Greater Bangalore Water Supply and Sewerage Project for Eight Cities in Bangalore Metropolitan Area, Project Development Report* (USAID 2003: xiv).

In this "Pooled Finance" framework, it was mandatory for citizens, named "users," to make a contribution toward the project:

> *User Contributions:* The Government of Karnataka has approved levying Rs. 8,500 per property as a one-time charge towards the water and sewerage project. We have assumed that this will be only for the water component. All urban local bodies have passed resolutions approving the same. Some of the local bodies have initiated collection of the charge. . . . However, to meet the project costs and based on discussions with GOK, it is proposed to increase the one-time capital contribution to Rs 8,500 and 17,000 for domestic and non-domestic customers.[7] While estimating revenues from this source, we have assumed that residential properties would pay Rs. 8500 per property and non-residential properties would pay Rs. 17,000.[8]

The user contribution was also levied for other related work such as "road cutting" to lay the pipes for water distribution:

> Moreover, an additional amount of Rs. 2,500 would need to be collected from properties towards road cutting charges of Rs. 800 and BWSSB connection charges at Rs. 1,700. The revenue from this portion needs to be transferred to these two agencies and not the project account. However, it needs be included in the user contribution in addition to Rs. 8,500 to make the project financially viable. This source is expected to generate Rs. 1194 million from FY 2003–04 through FY 2005–06.[9]

The National Institute of Urban Affairs (NIUA) described the Water Project as a project that exemplifies "teamwork" among multiple state and para-statal agencies, international donor agencies, and NGOs as follows:

---

7. The user contribution for the residential units was reduced later and corresponded to the square footage of the dwelling unit. Also a major discussion occurred among various state and para-statal agencies regarding how much contribution could be and should be expected of users living in the slums, known as the "urban poor."

8. USAID (2003: xii).

9. Ibid.

Karnataka Urban Infrastructure Development and Finance Corporation (KIODFC): Development and financial structuring
Bangalore Water Supply and Sewerage Board (BWSSB): Project Implementation
Directorate of Municipal Administration (DMA): Municipal Reforms
Janaagraha: Structured Citizen Participation
United States Agency for International Development (USAID): Credit enhancement
Water and Sanitation Program (WSP), World Bank: Demand Assessment
International Finance Corporation (IFC): Operation and Maintenance (O&M) Advisory
Strategic: Bond Issue (Urban Finance 2005)

The Financial Institution for Reform and Expansion (USAID-FIRE) developed the original plan for the project given their experience with a smaller project in Alandur in the neighboring state of Tamil Nadu. Janaagraha was responsible for organizing the citizens' participation component with funding from the WSP, a subsidiary of the World Bank. IFC, on the other hand, was working on plausible ways to privatize the O&M segment after the completion of the project. As mentioned in the NIUA list above, BWSSB or the Water Board would supervise the project in terms of the engineering of the water supply and the sewerage sections, with KUIDFC handling and directing all the required funds for it. Interestingly the NIUA or any of the official presentations do not mention the CMCs or local governments in whose jurisdictional area the Water Project will take place; it only mentions the DMA, which supervises these local governments.

The Experiment

The multifaceted innovation in financing was echoed as a true public-private partnership, or PPP, in presentations made at conferences and to the public by any of the organizations involved. The various organizations working on the Water Project through their understanding of water and the market activated a dense transaction of meanings around several concepts: reforms, governance, markets, PPP, citizen participation, credit, and so on. The density of the various notions generated an ever-expanding nebula of meanings (especially regarding the inclusion of the urban poor in the project that I discuss in Chapter 5).

On the other hand, the project was rather (and ironically) fluid, especially in relation to its financing model. In their article "Piped Water Supply to Greater Bangalore: Putting the Cart before the Horse?" (2009), Malini Ranganathan, Lalitha Kamath, and Vinay Baindur have recalculated the project cost, which had increased beyond the original estimate. The authors argued that the rising cost of the project is symptomatic of the problems inherent in what they term "market-based reforms,"[10] which seldom take into account the specific questions of equity, entitlement, and existing urbanization patterns specific to Bangalore. Ranganathan, Kamath, and Baindur (2009) identified four interconnected arguments that explain why they consider the project as a case of placing the "cart before the horse": (1) the citizen contribution has not assured timely and adequate services; (2) the project is disengaged from the patterns of urbanization in Greater Bangalore (3) the project is managed centrally, leaving the local governance bodies powerless and leading to low political and public acceptance; and (4) the changes in the financing model have led to public distrust. The Water Project is part of a wider reform agenda that has been sweeping the developing world since the 1990s. Sylvy Jaglin (2002) has claimed that a significant departure has occurred in the realm of public water services since the 1990s in the wake of economic liberalization and state reforms in developing countries. Rather than thinking of the state as the developmental machine, international donor agencies advocate that the state be active only in establishing and facilitating private-sector participation in the provision of water. However, even the market reforms that supposedly framed the Water Project in many ways were not aligned with the reality of the area that they were working on.

I am specifically interested in the urbanization patterns and the centralized structure of the project that Ranganathan, Kamath, and Baindur identify as two of the major issues of the Water Project. In this regard, what struck me first is the project's peripheral location, which I argue lends a certain angle to a social experiment that may be unavailable in the center. Rather than creating an artificial split between the center and the periphery, I demonstrate the power imbalance between the two and the effect it has on how the project was introduced, ratified, and authorized.

Greater Bangalore then comprised seven CMCs, Bomanahalli, Byatarayanapura, Dasarahalli, Krishnarajapuram, Mahadevapura, Rajarajeshwarinagar, and Yelahanka, and one town municipal council (TMC), Kengeri. The ULBs were formed in 1997 by the government of Karnataka under the Seventy-Fourth Amendment Act and are supervised by the directorate of municipal administration (DMA). The Seventy-Fourth Amendment of the Indian Constitution was adapted in 1992 in an effort to transfer more power to the ULBs and the Village Panchayat and to decentralize administrative responsibilities along the hierarchy of the federal system. The "Statement of Objects and Reasons" of the Seventy-Fourth Amendment Act is as follows:

---

10. The authors draw on Bakker's (2005) idea of "market-based reforms," which they explain as "a mode of management that relies on the market to meet economic and resource/environmental sustainability goals" (Ranganathan, Kamath, and Baindur 2009: 54).

In many States local bodies have become weak and ineffective on account of a variety of reasons, including the failure to hold regular elections, prolonged supersessions [*sic*] and inadequate devolution of powers and functions. As a result, Urban Local Bodies are not able to perform effectively as vibrant democratic units of self-government.

  2. Having regard to these inadequacies, it is considered necessary that provisions relating to Urban Local Bodies are incorporated in the Constitution particularly for—

    (i) putting on a firmer footing the relationship between the State Government and the Urban Local Bodies with respect to—
    (a) the functions and taxation powers; and
    (b) arrangements for revenue sharing;
   (ii) Ensuring regular conduct of elections;
  (iii) ensuring timely elections in the case of supersession [*sic*]; and
   (iv) providing adequate representation for the weaker sections like Scheduled Castes, Scheduled Tribes and women.[11]

Theoretically, the separation between Bangalore and Greater Bangalore was jurisdictional; the Bangalore Mahanagar Palike (BMP) administered the city of Bangalore and the CMCs administered Greater Bangalore. However, the juridical distinction between the two is inconsequential in terms of the discursive location of the city in the public imagination. The identification of Bangalore as the "Silicon Valley of India" indubitably includes Greater Bangalore since most of the IT companies, such as Infosys, are situated in zones dedicated to the industry, which are mainly located in the CMCs. These dedicated zones, such as the International Technology Park Limited (ITPL), are located in the Mahadevapura CMC, and the Electronics City is located in the Bommanahalli CMC.

Greater Bangalore is the urban sprawl that has developed over the last twenty five years as Bangalore experienced an explosion in IT and IT enabled services (ITES) industries. Rather than the actual city of Bangalore, which is relatively constrained in size, it is the peripheral Greater Bangalore that largely shouldered this transformation, accommodating both new businesses and residential developments. In the Resident Welfare Association (RWA) meetings I attended, the discussions dwelled on the infrastructural lacuna of the area and the difficulties of everyday life associated with it.

---

11. The Constitution (Seventy-Fourth Amendment) Act, 1992, available at http://indiacode .nic.in/coiweb/amend/amend74.htm (accessed December 19, 2006).

A prominent blog called "Sights, Sounds and Smells from Bangalore: Tracking India's Silicon City's Growth Pangs" posted the following news report from the *Times of India*:

The seven CMCs on the city periphery are poor cousins of Bangalore, where residents are deprived of basic amenities: bumpy, unasphalted roads with huge stones popping all along the stretch, lack of drinking water supply, drainage facilities and public transportation. Residents, who have been constantly egging on the authorities to improve the conditions, often take to the streets, while a few get the work done by pooling resources.[12]

Ranganathan, Kamath, and Baindur (2009) have discussed the urban settlement patterns in the erstwhile Greater Bangalore area in detail, particularly in relation to the region's access to water. They write:

Alongside technology parks and real estate developments, quasi-legal subdivisions known as "revenue layouts" have also proliferated . . . the revenue layouts developed from the mid-1900s onwards with the transfer of agricultural land to real estate developers who subsequently subdivided the land and sold it to buyers without getting the legal approvals. That is, unless the layouts were formally converted to non-agricultural land (through payment of a conversion fee), they maintained an illegal status. . . . Residents living in revenue sites and slums rely almost completely on ground water for their drinking and domestic needs. (54–55)

What comes across in the above portrayal is a chaotic and haphazard transformation of the rural fringe into the peri-urban. The in-between space amid the previous rural and the emergent urban is an important topographical feature here. The ethnography on the Water Project I present here draws on work I conducted on the Mahadevapura CMC, where Janaagraha was specifically active, and because this CMC had a relatively higher concentration of slums, the choice was relevant. Let us first look at the middle-class perspective on the Water Project, where I primarily worked with an RWA. The RWAs are mandated to include members from every social class in a given area; in reality, however, RWAs are exclusively middle-class enclaves.

One such RWA is the one in Pai Layout. Mrs. Tesse Thomas, who was an active member of this RWA in the K. R. Puram CMC, introduced me to her

---

12. Available at http://bangalorebuzz.blogspot.com/2005/04/give-us-roads-water-cry-residents.html (accessed April 20, 2005).

neighbors and colleagues at one of their regular Sunday meetings. I had met Mrs. Thomas earlier at a workshop organized by Janaagraha to introduce citizens to their concept of citizen participation in the Water Project. The Pai Layout RWA, along with the other RWAs of the area, was having its monthly meeting with the CMC commissioner and later joined the Janaagraha workshop being held in the same building.[13] After an initial introduction that day, Mrs. Thomas agreed to take me on a reconnaissance trip around her CMC that weekend.

The following week, I found Mrs. Thomas waiting outside her house for me as planned on a late Saturday afternoon. As we started walking past the arterial roads, we found ourselves negotiating mud paths that were not asphalted, and the constant automobile traffic had infused the air with a permanent sheath of dust. Mrs. Thomas had prepared me in advance, and I covered my nose and mouth with the dupatta to keep myself from inhaling the dust.[14] All through our walk, Mrs. Thomas seldom spoke. Later as we reemerged on the asphalted street, she finally said, "Did I not tell you?" I nodded in agreement. A visit to the area reveals the haphazard growth indicative of the city's unpreparedness to deal with the sudden economic boom when it transformed from the quiet "Pensioner's Paradise" to the frenzied "Silicon Valley." Mrs. Thomas pensively stated, "What you see is only the surface. The bigger problem is water. There is no water here; we have to buy water or dig bore wells."

Being a relatively new area of urban growth, Greater Bangalore was then not incorporated in the piped water supply and sewerage network provided by BWSSB in the juridical limits of BMP. In the middle- and upper-middle-class areas, the only access to water was through bore wells that were dug for individual buildings and then distributed through motorized pumps to faucets in the kitchen and bathrooms. In the hot summer months when the water table drops, water is usually purchased from water tankers sold by private suppliers at a higher cost.[15] In the slums, the CMC administration provides water through community standpipes connected to a motorized bore well usually twice every week. Given their economic constraints, the unreliable water supply from the CMCs and the whim of the valve man who operates the standpipes,[16] the slum dwellers usually walk on average 5 kilometers

---

13. Legal provision Seventy-Fourth Amendment.

14. A dupatta is a long scarf that is usually worn around the neck with the traditional long shirt, called kameez, and slacks, known as salwar.

15. Discussions around the water table were prevalent in every site that I worked, and most of them were based on the alarming rate at which the water table is receding in Greater Bangalore given the unrestrained digging of bore wells by residents.

16. The valve man is a vital figure in the water supply system in the slums. Theoretically he is expected to supply water to the slums under his jurisdiction on a rotational basis. However, it often turns into a personal agenda where he supplies more water to a particular slum in return for a favor, mostly monetary, or withholds water for an extended period if his expectations are not met or if he is challenged in some way by one of the slum residents. I elaborate on this in Chapter 5.

(2.2 miles) to access water from neighboring standpipes provided by BWSSB in the adjacent Bangalore metropolitan area.

A Paradox

The above is a dismal but at best only partial portrayal of Greater Bangalore. Greater Bangalore, as I mentioned above, is also home to the IT industry in developments like the ITPL and the Electronics City, which are equipped with state-of-the-art facilities in which to conduct business halfway across the globe. The architectural facade of glass and fine concrete buildings with running water in the dining areas and the bathrooms, green sprawling lawns with in-ground sprinklers, swimming pools, and so on, of the IT parks amplify and stage corporate affluence. The Water Board supplies water to the IT parks through dedicated water lines. Water is available, to use the administrative phrase, "24/7," which masks the issue of water scarcity outside the walls of these technology parks. Ironically, in the restrooms, Infosys displays little stickers near the washbasins and behind the stall doors to remind the user about the value of water and the urgency to conserve it. Both ITPL and Electronics City invest heavily in enhancing the natural aesthetics that include a variety of plants and trees, manicured lawns, and artificial lakes that constantly require vast amounts of water. Similar to the restrooms, little placards are placed along the green lawns and at the base of plants stating the importance of a balanced ecological existence.[17] It is not difficult to see that these paraphernalia are more a matter of surface CSR and capitalist niceties than about the environment.

The availability of water, entered as "Water Storage" on the ITPL website, is also illuminating. Note that the section is titled "Water Storage" rather than "Water Supply." While not directly acknowledging the issue of water shortage, storing more than enough water supplied through a dedicated pipeline from the Water Board makes an indirect allusion: "Water supply is through a dedicated 600 m pipeline from the Bangalore Water Supply & Sewerage Board, easily catering to a monthly consumption of 10,500 kilo ltrs. The main water storage tank holds a capacity equal to a week's consumption at the Park. Water quality is regularly checked, in compliance with WHO standards thus ensuring clean, healthy water all the time."[18]

---

17. A stretch of land located on the rear side of Building 1 of Infosys is dedicated solely to trees that are planted by official dignitaries who visit the place. A plaque bears the name of the planter and his official designation. While this is a gesture through which Infosys intends to honor their important visitors where they leave behind a mark, the plant as a "natural" marker supposedly also conveys the company's sensitive attitude toward the environment.

18. Available at http://www.intltechpark.com/infrastructure/water_storage.htm (accessed March 22, 2006).

The residents of Greater Bangalore have a mixed reaction to the presence
of IT companies in their area. Their optimism lies in the fact that the presence of the high-profile companies will draw the attention of the state to the
area where there has been a recent initiative of investing in high-quality infrastructure: "It is only the presence of ITPL in the area, that we can hope for
something to happen to us, as well; maybe the government will do something
to supply water to us as they do to these big companies here." On the other
hand, the feeling of being bypassed by the state is also evident. At the RWA
meetings, the comparison was drawn by citizens between the condition of
their infrastructure vis-à-vis that of the "IT companies just a few kilometers
away." The usual expression was: "How do they have everything?"

I mentioned this disparity to A. Krishnappa, the local member of the
Legislative Assembly (MLA) of the K. R. Puram and Mahadevapura CMCs.
He insisted that I recognize the "prestige these IT companies bring to our
CMC." He talked about a recent visit of a delegation from Singapore to his
CMC and the prospect it holds for his electorate: "We must be proud of ITPL
and the IT companies in Mahadevapura, they attract foreigners, and we want
to present them with a good picture of what we can achieve." He was referring
to a state visit by the Singapore prime minister, Mr. Goh Chok Tong, to meet
the then Indian prime minister P. V. Narasimha Rao. Understandably Krishnappa was honored that it was from his CMC that the idea of transforming
Bangalore originally materialized. However, that the Singapore intervention
transformed only pockets of the CMC remains a different story.

Around the time I met Krishnappa, Mahadevapura CMC was building its own website where citizens could access information and download
relevant forms to carry out various transactions with the state. Krishnappa
mentioned the forthcoming website to me as a sign of "progress" for a CMC
that is otherwise thought of as "backward," in a way signaling inadvertently
that he too acknowledges the paradox inflicting the area. The website was being developed by the eGovernments Foundation started by Nandan Nilekani
and Srikanth Nadhamuni "with a goal of creating an eGovernance system to
improve the functioning of City Municipalities leading to better delivery of
services to their citizens."[19]

I visited the website the day it was formally launched: It had an elaborate description of the ITPL along with pictures. At a later date when I met
Krishnappa at a public meeting in Mahadevapura, he wanted my opinion on
the website, which, according to him, "features the best we have." However,
my original question still remained unanswered. It specifically intrigued me
because given Krishnappa's democratic mandate as an elected councilor, his
presentation of his polity seemed deliberately partial. What were the contours

---

19. eGovernments Foundation, "About Us," available at http://www.egovernments.org/about
us.htm (accessed March 22, 2006).

of his politics that straddle democracy and the market, particularly in the context of water, which is not only a public good but also a decisive issue in vote bargaining, especially in the slums?

Chikka Venkatappa, the commissioner, who arranged my meeting with Krishnappa, agreed to the latter's portrayal of Mahadevapura as a region to be proud of despite the infrastructural problems of ordinary citizens who live there. Further, ignoring incidents of citizen protests that Chikka Venkatappa faces almost on a regular basis at his office (I witnessed several), the ITPL in his CMC was something he *had* to honor as a global emblem of success. Before I could broach the subject, Krishnappa himself had raised the upcoming Water Project. Like ITPL, he was excited about the project because it would "take care of the long-term water woes of my people, and I have been able to do something for them, as you can see, and not just for the IT people."

The change in his choice of words as he moved from ITPL to the Water Project is quite remarkable: while the ITPL is exclusive, the Water Project is plebeian—it is about "my people." As his democratic concerns surfaced during our conversation, his demeanor shifted from unqualified enthusiasm (regarding IT companies in ITPL) to an expression of optimism tempered with some hesitation. In a way to almost assuage his own anxiety, Krishnappa told me that he completely understood the magnitude of the fiscal responsibility the Water Project had placed on his administration. Failing to meet it would be detrimental to his political career and "to my electorate too because if water is now priced, it will be difficult for me to ask people to vote next time."

Krishnappa's comment about water as an instrument of electoral negotiation is a well-known campaign strategy in India, relevant mostly with slum residents. At Janaagraha and Infosys alike, politicians like Krishnappa were routinely indicted for their "greed" and "corrupt" practices. They argued that the introduction of the market, in the guise of a private operator independent of petty politics, who, unlike Krishnappa, is independent of electoral politics, would end such corruption. What is of interest here is not only how the market is legitimized but, more importantly, how the market gets interwoven with democracy. The following excerpt from the Janaagraha website about citizen participation in the Water Project clarifies this further:

> Throughout the world, the formal participation of "user groups" is increasingly being seen as one of the critical success factors for a range of public good provisioning. The central issue that occupies policymakers and service providers is not whether to bring user groups into the decision-making process, but how to do this in an organized and institutional manner. One additional element of this challenge is when the "user" is also a "citizen," where the context is not just the delivery of good quality services, but a larger one of effective func-

tioning of democratic institutions through participatory processes of decision-making.[20]

Here we encounter the market as a tool for democracy rather than development. Yet, as we can see, the market ideal circumvents and even disparages the democratic process itself. When I raised this issue, senior volunteers at Janaagraha explained that they were not necessarily against any particular councilor or politician but they felt their aim was to "sanitize" the democratic system by implanting the private operator, their shorthand for market, as an apolitical tool. The market, they reasoned, will make elected councilors, aspiring politicians, and administrators, as well as citizens, more accountable, especially in the use of public funds. "We do not want to do away with elections for sure," one of the volunteers assured me. "We just want to let those politicians know that as citizens we are now watching them." Janaagraha's interest in the Water Project was not about simply creating a water market. It was mainly intended to serve as a paradigmatic example of how to make the state accountable and effectively commodify amenities and services.

Indisputably, Krishnappa's uneasiness with the Water Project is related to his trajectory as a politician who has amassed significant wealth. His splendid house, where I was invited and treated to a sumptuous meal of *dosa* and *sambar*, and the fleet of expensive cars waiting in his front porch are clear indicators of his "corrupt" practices. However, to look at corruption specifically as an issue that concerns politicians alone is misplaced. It deflects us from similar practices in other places, such as appointed public servants and corporations. The legitimacy of the market is thus based on a construction of "corruption" that is not only limited but also naturalized as occurring only in specific sectors of the society.

To elicit his response, I mentioned to Krishnappa, that perhaps the Water Project had little to do with water supply because its advocates are largely publicizing it as an experiment to end corruption. Not surprisingly, he defended himself but at the same time reminded me that the Water Project "may end up digging graves for its own backers because if they think they can cover up the things they did, they are wrong." Krishnappa's assistant came in to remind him of his next appointment, so our conversation ended at that point. As I was walking out the door, he called me back. In the tone of a veteran politician he said, "Madam, you may come up with a whole new research then."[21]

Chikka Venkatappa picked me up. While driving back to his office, he wanted to know if the meeting was useful. I said, "Yes, but I was a little perplexed. Citizens in your CMC seem very upset about the infrastructural problems which Mr. Krishnappa did not seem to recognize till we talked

20. Available at http://www.janaagraha.org/node/113 (accessed March 24, 2006).
21. Women in India are referred to as "madam" as a way of showing deference.

about GBWASP." He responded, "He is a great man. Water is not his problem, madam; it's mine. I have a tougher job." As my fieldwork progressed, it
became clear that it is indeed the commissioner who is responsible for infrastructural systems and also placating disgruntled citizens, especially the slum
residents, through vote bargaining.

Fluid Power

Peri-urban spaces that frame global cities especially in the developing world
are marked by uncertainty. The uncertainty of the periphery is twofold.
First, it is perceived as a failing space, and, second, the failing justifies it
as a space of experimentation. Bangalore, for example, is radially expand-
ing outward where the IT companies and the related lifestyle places, such
as gated residences, foreign car dealers, shopping malls, and so forth, are
located. The adjacency of luxurious buildings alongside derelict structures
is not only symptomatic of a new form of inequality but also reinforces the
sense of uncertainty. While large-scale real estate developers were able to get
their plans sanctioned, independent homeowners were struggling with such
administrative procedures. They were also worried, as was often discussed in
the RWA meetings, about the declining value of their properties. Thus it not
surprising that while the Greater Bangalore area received state attention, the
attention was selective in terms of infrastructural investment.

The IT parks and gated residential enclaves enjoyed dedicated water pipes
and bulk water supply from the Water Board at a substantially higher cost.
Highway billboards and Internet advertisements for upcoming residential developments in Greater Bangalore, especially near the ITPL and the Electronics
City, list all the "world-class" facilities they offer, such as supermarkets, gyms,
swimming pools, clubhouses, walking trails, and so on, but water is seldom
mentioned. The following is from one of the Puravankara Projects, a leading
real estate developer:

> Purva Venezia is inspired by the magical landscape of Venice. From
> the sparkling waterways with gently rocking gondolas. . . . Every as-
> pect of this high end home is thoughtfully designed—be it the vitrified
> flooring, the ornate lobby or even the water purification plant. . . . Your
> home will be so self-contained that you may never have to step out for
> days on end. A supermarket, a gym, a swimming pool, a health spa,
> a Jacuzzi, a tennis court and many more. If you've always believed in
> high quality living, it's time to put your beliefs into action.[22]

---

22. Available at http://www.puravankara.com/projects/venezia/venezia.asp (accessed
April 17, 2006).

The use of "waterways" makes water a seemingly abundant resource, and "water purification plant" indicates health and well-being. Overall, there is an evident de-emphasis on water itself; it is a given. The bulk water received by the gated apartment communities is stored in large overhead tanks in the terraces (often invisible) and supplied to the residents consistently. Thus the simulation of a regular water supply buries the existing shortage. Given their price range, which hovers around rupees 45 lakhs (approximately US$75,000), and the kind of lifestyle they promise, these new residential complexes are primarily intended for young professionals, primarily working in IT. The cost of installing a dedicated pipe system and the purchasing of bulk water is worked into the price of the apartments, but it remains unmentioned and undisputed. Most Infoscions who either lived in these gated communities or were buying properties in Greater Bangalore were unaware of the upcoming Water Project. They were immune to such infrastructural problems given the money they had invested to bypass the average service level of the public water supply system.

My work with the Water Board revealed that while dedicated water lines were technically sanctioned against a stipulated higher payment, it was also approved on the basis of financial reliability, which mainly included the IT companies and prominent real estate developers. This evaluative means, which went beyond formal structures, opened up an illegible and likewise an illegal space of transaction between the state and the citizens where claims were made, contested, approved, or denied.

Greater Bangalore as a peripheral location is not unique. The illegibility of peri-urban spaces is further exacerbated by unclear jurisdiction. Neither the state administrators nor the citizens had a clear idea about the jurisdictional limits of Greater Bangalore—whether it was the BMP or the CMC. On the other hand, in my findings the theoretical basis of jurisdiction seemed fairly simple. A given area was upgraded from a Village Panchayat to a TMC, and finally to a CMC based on increase in population and hence the need for a larger-scale administrative apparatus. An officer with the Directorate of Municipal Administration (DMA) that oversees the working of the Greater Bangalore area advised me to look at their mandate of work, which traces this gradual jurisdictional transformation of an area as follows:

> The Government of Karnataka has reconstituted the municipalities according to the 74th Constitutional Amendment Act. The towns have now been classified based on the population and other criteria as Town Panchayat (Population 10,000–20,000), Town Municipal Councils (Population 20,000–50,000) City Municipal Councils (Population 50,000–3,000,000) and City Corporations (Population 3.0 Lakhs and above).[23]

---

23. Available at http://municipaladmn.kar.nic.in/aboutuse.htm (accessed April 17, 2006).

However, in reality the land jurisdiction between the BDA and the CMC regarding who is authorized to confer land rights and sanction real estate initiatives was nebulous. Often citizens complained that they never got a clear picture of the various certificates they needed to produce to the government agency to secure a permit to build houses, let alone which agency they should approach, the BDA or the CMC. This was, as they said, a problem with "ordinary honest citizens," but the "big builders" usually played around with this kind of ambiguity "to secure their permit quickly and easily."[24] There are also administrative problems regarding the collection of taxes and betterment charges. The real estate section of *India Times*, a prominent news media website, cautions prospective buyers who were considering investing in the CMC areas:

> Presently, CMCs have stopped collecting betterment charges and are not issuing Khatas.[25] Owners of sites are filing SAS for paying property tax but in the absence of Khata, the building plans cannot be approved.[26] Until the government regularises these properties and brings in some regulations to put things in order, it is better to keep off such properties."[27]

Often members of the RWAs who worked in other sectors mentioned to me how they invested their lifelong savings to build a house they had always desired only to find that the taps did not have water. I was curious to find out why an area that is routinely talked about negatively in terms of its appalling infrastructure and ambiguous jurisdiction became a site where people invested in private property. Were they not aware of these problems? On the

---

24. By "big builders" they mean the real estate companies that are now buying off vast plots of land to build large-scale expensive and luxurious gated communities self-sufficient with all infrastructures. The roads that lead from the city to these CMCs are lined with billboards that these building companies use for advertisements. The advertisements typically feature a single-family house similar to a suburban house in the United States complete with green grass front yards. Quite often the smiling faces featured in these advertisements, like a parent-child duo, are white-skinned and blond-haired, promoted as the image of a good and desirable life.

25. A "Khata" is a certificate issued by the revenue department of the state, which can be obtained by the owner of a property on the payment of twenty-five rupees ($0.50) provided that all the taxes due on the property have been paid.

26. SAS stands for Self-Assessment Scheme, a way of paying property taxes where owners evaluate their property and report it for the corresponding tax to be levied on it. In 2004 it was proposed that SAS be replaced by Capital Value System (CVS), where revenue inspectors will assess and levy the taxes. CVS was opposed from many quarters for the fear that it will reintroduce the "Inspector Raj." Janaagraha organized a campaign called "Taxation with Transparency" opposing the CVS. It argued in a signature collection drive: "Did you know? 50% of Bangalore pays for the other 50%. Bangalore needs more Taxpayers & not more taxes."

27. Available at http://realestate.indiatimes.com/articleshow/1344773.cms (accessed April 17, 2006).

contrary, as I discovered, most people who decided to invest in real estate in Greater Bangalore knew that there would be no Water Board supply and the only way they could obtain water would be through bore wells for which they had to pay the contractor a separate fee during or after the construction of the house or through tanker supply during the summer. Engineers at BWSSB reiterated the same, claiming, "Whoever invested in Greater Bangalore knew that we do not supply water there except for bulk water."[28] Instances such as these add to the everyday uncertainty in governance regimes and are also symptomatic of the ambiguity of the urban terrain I am discussing here.

Though powers were devolved to the local bodies, the effectiveness of the decentralization and the implementation of the Seventy-Fourth Amendment were debated from the start. Thus the allocation of jurisdictional and administrative power remained undecided between the CMCs and the Bangalore City Commission (BCC). The Seventy-Fourth Amendment in reality generated a number of administrative gaps that relegated areas such as Greater Bangalore to a murky zone. However, citizens in the middle-class neighborhoods argued that the murkiness, though a difficulty for them, was productive for administrators and politicians in the area. In another sense, it enabled corruption. Corruption here is thus understood as a derivative of a nebulous governance structure lacking accountability.

As Akhil Gupta (2005) has argued, the narrative of corruption is a defining component of social life in India and these narratives have a fairly definite structure and elements. In the case of Greater Bangalore, the jurisdictional murk translates into an "ethical murk" where state administrators and politicians can and do suspend standards of "good" governance with relatively greater ease. In 2003 the Committee for Redressal of Citizens Grievances, known as the *Lokayukta* of the government of Karnataka, exposed a scam of nearly rupees 231 crores (US$6 million) of misappropriation of state funds in the seven CMCs.[29] The then chief minister, S. M. Krishna, acting on the advice of the *Lokayukta*, suspended nineteen CMC officials.[30] Greater Bangalore was thus conceived as a dual space of economic success (the presence of the IT industries), on one hand, and ethical bankruptcy, on the other.

I argue that the Water Project was both an administrative rationale and a political device to legitimize the project as an effort to "sanitize" the area.

---

28. BWSSB supplies bulk water to business establishments such as ITPL and the Electronics City and is usually at a higher cost than the regular rate. Some residential areas in Greater Bangalore also receive bulk water supply from BWSSB, for example, a few wards in the Mahadevapura CMC. However, residents consider this as an interim solution that they expect will be replaced by the Water Project.

29. The idea of the *Lokpal* at the center and *Lokayuktas* at the state levels was introduced to address citizens' grievances and enhance the functioning of the administrative apparatus. The Karnataka State Legislature enacted the Karnataka Lokayukta Act in 1984.

30. *The Hindu*, December 11, 2003, available at http://www.hindu.com/2003/12/11/stories/2003121104080400.htm (accessed November 21, 2006).

The Water Board, which traditionally supplied water in the BMP area, was entrusted with a project in the CMC area over which technically it did not have jurisdictional mandate at that time. I persistently inquired with the administrators at the Mahadevapura and K. R. Puram CMCs if they could have implemented the Water Project independent of the Water Board, but I never received a definite answer. The answers were similar to what Chikka Venkatappa once mentioned, "It's better for them to do it." When I asked why, he responded with his usual smile that over time I had come to interpret as "You know why."

On the other hand, middle-class residents of Greater Bangalore deeply believed that this ambiguity would be resolved once the Water Project is completed and Water Board starts supplying water to their area like they do in the BMP area. Water was perceived as the leveler of the uncertainty that continues to define Greater Bangalore as away from yet a part of Bangalore. The Water Project also redefines the politics of the region, as I show below, in terms of the relatively lower position of the seven CMCs and one TMC of Greater Bangalore in the governmental hierarchy.

The memory of the scam I mentioned above was vivid, and it was regularly evoked by citizens to describe the inept administrative apparatus. At the governance level, the scam is seen as a failure of decentralization and raised questions about "fiscal discipline." In an attempt to resolve its uneasiness with the supposed "corruption" and "inefficiency" of the CMCs, the government of Karnataka initiated the plan to merge Greater Bangalore (along with 111 villages) with the city of Bangalore under the BMP, which was eventually implemented in April 2007, making it the Bruhat Bengaluru Mahanagara Palike (BBMP). The proposed merger was tantamount to the reversal of the Seventy-Fourth Amendment. Most middle-class citizens I knew in Greater Bangalore, who saw it as their only exit out of a corrupt system, supported the merger.

However, several critics deplored the merger. Vinod Vyasalu, executive director of a policy analysis organization called the Center for Budget and Policy Studies (CBPS), contested the merger in an article published in the *Deccan Herald*. He interrogated the basis on which this merger was being vindicated: "It is assumed that the BMP works better. No study has been undertaken of the working of the BMP, and its relative performance compared to the CMCs. A few citizen initiatives have shown many weaknesses in the BMP. . . . On the assumption that the CMCs cannot deliver the goods, and that the BMP is better positioned to do, such a major change is being brought about."[31]

---

31. *Deccan Herald* (Bangalore), March 7, 2006, available at http://64.233.161.104/search?q=cache:ZntfqFPOmSUJ:59.92.116.99/website/DOCPOST/mar06/HD11-of-mergers-and-demergers.pdf+vinod+vyasalu+CMCs&hl=en&gl=us&ct=clnk&cd=1 (accessed November 21, 2006).

The hierarchy between the BMP and the CMC is evident in Vyasalu's analysis, and the CMC is divested of its power based on a set of assumptions. The performance of the BMP is deemed impeccable while that of the CMCs is outrightly condemned.

Vyasalu and Sharadini Rath, a project leader at CBPS, expressed this blatant domination of the state government during one of our conversations. At that time, CBPS had just completed a study of the financial system of the CMCs that revealed a different image of the councilors. As elected representatives of the people, several councilors were dedicated to their wards, but they were strictly constrained in several ways: First, they lacked political power to veto any decision of the government of Karnataka imposed on them; second, they were deficient in funds to implement their own projects; and, finally, they were understaffed in terms of human resources. The CMCs largely were not deemed as "governments" but as corrupt bodies in need of constant control and supervision. Sharadini asked, "Do they mean to say that only councilors in the CMC areas are corrupt and none out of the one hundred in the BMP area are? Does this also mean that none of the Indian Administrative Service officers are corrupt if they only want to indict the Karnataka Administrative Service officers?"[32] It is plausible that the subordinate position of the CMCs make them relatively more vulnerable to accusations that may be overlooked in the BMP area. Nevertheless, I wondered at the supposed political naiveté of the CMC politicians and administrators if one were to accept Vyasalu's and Rath's analyses. My ethnographic experience with the area suggests that Greater Bangalore is far more complex than a neat power differential between the BMP and the CMC political and administrative systems.

Why is Greater Bangalore important to BMP? Is it merely about augmenting its power, or is it about extending a certain *kind* of power over an area that is open to new social designs? How are the IT companies productive for the BMP administration? What are the legitimate bases through which diverse populations lay their claims on the land, water, the infrastructure, and the state? Is the periphery critical to the IT dream of a new Bangalore that can be showcased as a world-class city to the world? I also wondered if this is a variation of colonialism, one experientially different from what India experienced for more than three centuries. Does it draw on the genealogy of colonialism and recast it in a different form?

---

32. I do not mention the dates of other interviews. IAS refers to the Indian Administrative Service and KAS to the Karnataka Administrative Service. The IAS officers are recruited through a national-level test and are considered superior to the KAS officers who qualify in the state-level test conducted by the government of Karnataka. Correspondingly, given their national scope, IAS officers are placed in relatively prominent administrative jobs compared with the KAS officers.

Land and Water

While Mrs. Thomas and her neighbors avow their rights as citizens to basic infrastructure, it is not quite equivalent to the ways in which IT companies can lay claims on the state. This is surely a way to think of power and access to the state. However, here I also propose to extend the story of power difference to ask what kinds of imaginations a peripheral location like Greater Bangalore enables and how IT is instrumental in it. Thus, I often inquired if Greater Bangalore would develop independently of IT.

The answers that I received from RWA members, administrators, and politicians are in the negative. While everybody disputed the notion that Greater Bangalore is tantamount to IT, they certainly think, as one bureaucrat surmised, that "a lot of Greater Bangalore is because of IT; the city cannot take in this [meaning IT] kind of growth." The bureaucrat continued, "We have to move outwards; Greater Bangalore faced a population growth and the government had to start doing some work in that area. Greater Bangalore happened because IT happened to us."[33]

The kind of dynamics the presence of IT in the Greater Bangalore area presents is intricately tied with the way the area was perceived as a zone of new possibilities, of new lives, a new individual, a new social, all of which emerge out of a new successful global industry and its concomitant aspirations, especially of owning a house. Since the 1990s, the banking sector has undergone considerable change to support liberalization, that is, to make money readily available through the newly introduced credit system for high-end purchases such as houses and cars along with credit cards. This opportunity was unprecedented for the middle class, who in the past depended mainly on savings from life-time earnings. It was usually after or close to retirement that one could afford a house, for example. The available credit and the choices offered by the market rearranged and certainly augmented the consumption practices of the middle class: they accelerated the urge to own and emptied credit of its earlier social stigma. Interestingly, credit cards acquired a social standing, while cash was seen as interfering with the speed of the global economy.

Water is curiously linked to this sprint to globalize. Landownership is directly allied with claims to water. When I spoke to residents about their problems with water supply, the subject of landownership was inevitable. Most complained that they had bought either a piece of land or real estate in the Greater Bangalore area thinking that the administration would be providing water, which is "after all a basic amenity." Land and water form a crucial amalgam in the legal discourse of governance and citizenship. The BWSSB

---

33. Conversation with an administrator at the KUIDFC, November 2004.

Act of 1964 defines the "occupier" and the "owner" of a building who are entitled to receive water, respectively, as follows:

> (13) "occupier'" includes—
>   (a) any person who for the time being is paying or is liable to pay to the owner the rent or any portion of the rent of the land or building in respect of which such rent is paid or is payable.
>
> (14) "owner" includes a person who for the time being is receiving or is entitled to receive, the rent of any land or building whether on his own account or on account of himself and others as an agent, trustee, guardian or receiver for any other person or who should so receive the rent or be entitled to receive it if the land or building or part thereof were let to a tenant.[34]

The legal category of the "occupier" is closely coupled with that of the "owner" of the land. Even if the property is a rental, ownership is the prime criterion for BWSSB to supply water. In the affidavit that citizens have to provide to BWSSB when applying for a new connection, the first oath explicitly states:

> 1. I do hereby declare that I am the owner of the house premises built on site _____ Bangalore . . .
> 4. I hereby declare that there is no other claimants in respect of the said property.[35]

However, this does not imply that BWSSB does not provide water to citizens who do not own land and/or live on rental properties or fail to provide evidence of ownership or reside on disputed land. The seventh oath of the same affidavit apprehends the possibility of such a situation after BWSSB has installed water connection under these circumstances:

> 7. I shall also agree to take the responsibility of solving dispute if any by person or neighbor BCC[36]/BDA/regarding water supply connection and I solve myself without involving BWSSB into litigation's and if any the necessary differences charges shall be paid by myself to BWSSB without fail.[37]

---

34. Government of Karnataka, Department of Law and Parliamentary Affairs, Karnataka Act No. 36 of 1964. The Bangalore Water Supply and Sewerage Act 1964 and the Bangalore Water Supply and Sewerage Rules 1964 and Regulations, Bangalore 1984.
35. Affidavit for BWSSB water supply connection given to me by one of the BWSSB engineers.
36. Bangalore City Corporation.
37. Affidavit for BWSSB Water Supply C.

As engineers in several subdivisions of the Water Board mentioned to me, that water is disconnected immediately in case any land dispute crops up.[38] However, in my discussion of water problems with Mrs. Thomas and her neighbors in the Pai Layout RWA, the issue of disconnection never arose. But my clarification question, "Does BWSSB have the mandate to disconnect water supply?," was not well received. Not only did frowns appear immediately but everyone also appeared surprised that, being part of the middle class, I would even pose such a question. The president of the RWA rose to the occasion, retorting, "Why would they? We live in a respectable neighborhood, that stuff they [Water Board] do only in the slums. We always pay our bills. That's not even something to talk about, and we all bought our lands and houses with money that we have earned with much honesty all our lives."

Defaulting on the payment to BWSSB or living on disputed land is considered both an aberration and also morally reprehensible among the middle class. The particular reaction to my question is not limited to merely to the financial ability of the middle class alone. The possibility of defaulting on a payment or living on disputed land is considered unthinkable. The commitment to make payments on time and living on land carrying explicit entitlement rights are indicators of honesty, a form of life that is placed beyond ethical breakdowns. As Raymond Williams reminds us, feelings far from being haphazard are rooted in deeper structures; they are the "complex relation of differentiated structures of feeling to differentiated classes." Following Williams's "'structure' as a set," values of honesty and ownership that emerge from it interlock with the materiality of land and water (Williams 1977).

"Break the Cycle"

In one of my interviews with him, K. P. Krishnan, then the managing director of Karnataka Urban Infrastructure Development and Finance Board (KUIDFC), located the Water Project at the intersection of "innovative financing," "municipal reforms," and "structured citizen participation." All these, he anticipated, would "break the cycle of corruption of the state and the expectations of the citizens." Besides being an avid supporter of Janaagraha himself, as a public servant he also emphasized the importance of an MoU between the government of Karnataka and Janaagraha. He emphatically held that "citizens need to take responsibility of what is being done for them." While citizen participation in state and nonstate projects abound in India, the signing of a MoU between an NGO and the state is unique. The MoU (pending at that time) significantly animated discussions of citizen participation in Janaagraha and was always mentioned with a sense of pride to newcomers.

---

38. The Water Board engineers would remind me time and again that the process of disconnecting water supply is prevalent more in the slums than in the middle-class neighborhoods.

For Janaagraha the MoU was a vital step in their attempt to change public governance: the MoU would formally and publicly recognize citizens as equal "stakeholders" in the Water Project. Even before it was formally signed on May 21, 2005,[39] the MoU had generated a sense of achievement; however, it also evoked anxiety regarding Janaagraha's legitimacy as an organization to invite citizens to participate in the project. The anxiety was often enunciated in the meetings as, "The MoU has not yet been signed" by the members of the Water Project team typically as a reminder to one another of their still uncertain status with the government of Karnataka and the possibility of being questioned by citizens of their authority to initiate "official" citizen participation. Participation in this instance did require citizens to pay the user contribution, a relatively new concept in urban infrastructural work in India.

When I looked for a precedence, the only other instance I discovered in which such a contribution was solicited was for a sewerage project in a small town called Alandur in Tamil Nadu, adjacent to Chennai:

> *Deposit mobilization* [sic]: The municipality proposed to collect connection deposits of Rs. 5,000/per household and Rs. 10,000/per non-household (non-domestic/commercial) for providing sewer connections. By the end of year 2004–05, the municipality proposed to provide connections to at least 21,869 households and 560 non-household entities, which would entail a collection of INR 1,150 Lakhs.[40]

While talking to Anand Jalakam and Vijay Padbmanabhan, former consultants of Financial Institutions Reform and Expansion Project (FIRE-D) who worked on the Alandur project, I wanted to know the overlaps and the differences between the two. While the Alandur project could be seen, they explained, as a precursor to the Water Project in terms of the user contribution, it was much smaller in scale. While the Water Project will cover eight ULBs, Alandur is one such local body and consequently the population covered was significantly small. Further, the Water Project involved a multitude of players located nationally and internationally. However, both Jalakam and Padbmanabhan linked the two projects to the notion of reforming governance.

These differences aside, I was especially interested, first, in the user contribution, which is historically unprecedented in public infrastructure work

---

39. Janaagraha withdrew from the MoU in early 2006, stating lack of transparency and willingness of the state to address the issue of the urban poor as the prime reasons. When I returned to Bangalore in 2006, volunteers at Janaagraha reiterated the above reasons to me. However, other NGOs in the city, who were already wary of Janaagraha's role as enabling water privatization at the cost of urban poor, saw these reasons as a way out of the project in the face of mounting criticism.

40. *The Alandur Underground Sewerage Project: Experiences with Implementing a Private Sector Participatory Project, Final Report*, National Institute of Urban Affairs (NIUA) and Indo-US FIRE (D) Project.

in India. Second, as I mention above, why did Alandur not stir the kind of experimental idea like it did for the Water Project? Furthermore, it took me a while before I was able to locate the Alandur project when it was the only precedence available in user contribution.[41] I do not think this is a mere oversight. Instead, it encourages us to raise a different question: What is special about the Water Project if one were to put aside the argument of scale? What matters in the Water Project is not just the idea of user contribution; it is, more importantly, the rhetoric underlying the idea.

Citizens have traditionally paid for residential water supply in India, albeit at a reduced rate, which is thus not unprecedented. However, what is unprecedented is the experimental nature of the Water Project both in its "pooled" financing and the language of "stakeholders" deployed to legitimize it. As a neoliberal project, the state would not only incorporate the private sector in the task of governance but also start emulating its model (Scrase 2006). The pooled financing model brings the state, the citizens, and the market within the same frame of reference; it obscures the structural and substantive relations among them. Under the section "Advantages of Pooled Finance Framework over Traditional Institutional Finance" in the FIRE report, the following features are listed:

i. The pooled finance framework brings in fiscal discipline in municipal operations especially for debt servicing.
ii. The proposed pooled finance framework with built-in credit enhancement measures, including USAID DCA guarantee, will reduce the cost of borrowing for the ULBs compared to traditional sources of institutional finance.
iii. The proposed pooled finance framework will also obviate the need for a State Guarantee.
iv. Under this project, 35 percent of resources are mobilized from the users and debt accounts for only 41 percent. This reduces the burden on local bodies as well as the state government. In the traditional financing system, this may not have been possible, as urban local bodies tend to depend heavily on State Government for project finance in terms of debt servicing and contributions.
v. The pooled finance framework envisaged under this project will establish a precedence in the state and may lead to reduction of pressures on the State Government for resources for urban infrastructure and result in improved infrastructure through replication of similar projects.[42]

41. The Alandur project was mentioned by one of the team members in Janaagraha when the need to document the work of the Water Project team arose as a precedent.
42. FIRE Report Framework for the Water Project.

The urgency of municipal reforms in Greater Bangalore, reducing the burden on the state, increasing ownership of the individual "user," yielding space to the market forces, and enhancing the credibility of the state with international donors, form the critical matrix on which the Water Project was based. This method of funding an infrastructure project was normalized among the state officials at the Water Board, KUIDFC, volunteers at Janaagraha, and the citizens, particularly in the middle-class neighborhoods. Nobody seemed to question its unique and unprecedented nature. Every attempt I made to engage members of the Pai Layout on the issue of user contribution was truncated with the usual refrain: "We need a reliable supply of water." The user contribution did not disturb the few who did raise questions. Rather they were annoyed that the state was "asking for money again when we have already paid the Betterment Tax during construction." The CMC officials supposedly misappropriated the Betterment Tax paid by citizens, which was revealed in the scam that was exposed by the *Lokayukta*.

However, other NGOs in the city, who were particularly against privatization of basic amenities such as water, criticized Janaagraha for serving as a government accomplice to endorse privatization at the expense of the urban poor. The founder of one of the NGOs, who worked with slum children in Bangalore and Greater Bangalore, described what he called "the game of GBWASP": "They [the BMP government] want to make sure that they put an end to the CMCs, so they piled them with GBWASP, something clearly the CMCs cannot handle on their own. This way BMP can have power over the CMCs. And then they got a NGO, i.e., Janaagraha, to convince people that they should pay if they need water. It is all very simple!" I wanted to know: "Was the GBWASP not passed by the CMC councils?" He was silent for a while, then said, "Everything is passed on paper. There is no problem there; it lies elsewhere." He alluded to the fact that the manner in which the Water Project was approved conceals more than an administrative quandary.

Later, in a casual conversation with an administrative clerk at the Mahadevapura CMC office one afternoon, he recollected for me how the Water Project resolution was passed. "The council was meeting on some other issues; I do not quite remember what, ordinary stuff, I think. Then at the end the commissioner passed a sheet of paper around saying it was for a water project and that it is good for the CMC. He did not seem to know much. But when the council members heard 'water' they all agreed, and so the vote was taken."[43]

From what apparently seemed like a democratic process, where the "sheet of paper," I assumed, would have mentioned the fiscal onus about to be placed on the CMCs, I wanted to know: "Did the council then know that the CMC will be under debt and that it was also responsible for collecting the user

---

43. Conversation, December 2004.

contribution? And what is the proposed service level of water supply?" All the answers came back in the negative. The manner in which the resolution was passed for a project that is experimental and grandiose appeared dubious. Others in the CMC office also confirmed the abruptness with which the Water Project was introduced as simply about water supply and the haste with which it was finalized. The process of gathering consent among CMC officials was not properly followed; in addition, the complexity of the project was not even discussed. Instead, it was presented as necessary (since it is about water) and a routine addendum toward the end of the meeting. Later, a junior volunteer, who was also present during the CMC session that day, described the transaction more as an "imposition" on the councils by the DMA rather than a democratic process of debate to arrive at a consensus to accept the colossal project.

However, despite other NGOs' criticisms, there was some discomfort about the user contribution in Janaagraha. The Water Project team in particular was aware of the kind of accusation being leveled against them by other NGOs in the city. The team was cognizant that while they ardently believed in the market as a panacea for a "corrupt" state, the task of instituting the market as the principle of governance was arduous. The general consensus in Janaagraha was that although the government of Karnataka designed the Water Project, it was Janaagraha's ethical responsibility to offer a platform for citizens to participate and ensure that the state is accountable in this process. As Ramesh Ramanathan put it in one meeting, "If they [citizens] understand the issue [water shortage], they will come on their own; here is an opportunity to participate."[44] Ramanathan was referring to the pilot workshop that was being planned to evaluate the level of interest among residents of Greater Bangalore to participate in the Water Project.

This was also stated on the Janaagraha website:

> The central issue that occupies policy-makers and service providers is not whether to bring user groups into the decision-making process, but how to do this in an organized and institutional manner. One additional element of this challenge is when the "user" is also a "citizen," where the context is not just the delivery of good quality services, but a larger one of effective functioning of democratic institutions through participatory processes of decision-making. . . . The GBWASP is a project that offers a canvas of this magnitude, integrating all categories of stakeholders. Hence, viewed in one light, a broader objective and intent of GBWASP could be to ensure the organized and structured design of citizen participation in a manner

---

44. Water Project weekly meeting, October 18, 2004.

that acts as a template for future Urban Infrastructure projects in India and beyond.[45]

At the same time, it is important to remember that the idea of the platform was in another sense reasonably inert in terms of citizen participation. Janaagraha did not question the rationale of the state to collect user contributions above and beyond the Betterment Taxes. In citizens' meetings, the reassertion of Janaagraha's role as a platform was usually used to circumvent queries about the user contribution. Instead, the rhetoric simply oscillated between participants and stakeholders, as though their interchangeability was a nonissue. The boundaries or definition of neither were delineated. In this context, however, I disagree with some social critics in the city that money alone lured Janaagraha. It does not justify the intense debates I witnessed at their meetings, which often focused on how to present the rationale of the Water Project in citizens' meetings. It was quite clear that everybody recognized the market but had not yet come to terms with its newness vis-à-vis the established socialist-redistributive model. Rather, the struggle offered Janaagraha an opportunity to figure out and refine their stance on "this new thing called the market" that was realigning both the civil society and the state in a new relationship. What would be their role be as an organization aspiring to reform governance and citizenship? However, it is one thing to launch brand-new cars; it is another thing to institute a water market. Water needed a careful rhetoric, which, if necessary, should remain nebulous for a while. Moreover, Janaagraha was focused more on making *existing* state processes transparent to the citizen rather than questioning the basis of the Water Project. The project was a *given* and besides, to Janaagraha it was a prime opportunity to stage the value of the market in reforming urban governance and citizenship in practice.

I recalled Vinod Vyasalu and Sharadini Rath and wandered about the depth and extent of depoliticization the Water Project might generate. How does Janaagraha propose to reform governance by eluding the basic issues of power, between the nucleus and the periphery, between the state and the civil society? Where does freedom from the state promised by the liberalization of 1991 and upheld by the "success" of IT lie for ordinary citizens? While such notions of neoliberal governance now abound in the developing world, the Bangalore experience still offers a unique instance. How does one explain the nature of this freedom? Further, how do we relate the new form of governance with the idea of freedom? Where do we look to identify the basis of this freedom? In Chapter 3, I address these questions. I specifically explore software development process as a form of knowledge production that tracks,

45. Available at http://www.janaagraha.org/archives/jalamitra_participatory.htm (accessed July 1, 2005).

manages, and aligns information. I argue that this mode of organizing information is underscored by a modernist ontology where there is no scope for ambiguity. When IT entrepreneurs promote corporate governance to amend public governance, what we basically encounter is a governance paradigm that transforms the relationship between the state and its citizens as though it were working off a software program.

3

Software Nation

Indumathy, informally referred to as Indu, walked across the street through the light morning mist as I waited for the Infosys bus to arrive. I had not seen her on the bus the day before on my way back and wanted to know if she stayed back after usual work hours. She told me she had to "because the client changed the requirement in the last minute." Recollecting my days as a software developer, I empathized with her. This is common in the industry and is usually an added burden on the project team as the deadline approaches. I wanted to know if the client had signed off on the "Requirement Specifications," usually known as "specs." She explained, "Yes, they did, but you know they are clients and that too in the United States, so we have to do what they say." Indu did not look particularly happy, but, like other IT professionals, she accepted these "small troubles" because "the IT industry is otherwise a great opportunity."

I met Indu on my first day at the bus stop. She is from Chennai in the neighboring state of Tamil Nadu and moved to Bangalore after accepting a job offer from Infosys. Currently she is living with her aunt in Cox Town, which she and her family held was "a safe arrangement for a single woman in a new city." The Infosys bus arrived at 7:05 every morning at Banaswadi Road in Cox Town. Like others, I usually arrived at the stop ten minutes prior to the pickup time, which also gave me the opportunity to converse with my fellow commuters.

Infosys is located in the Electronics City, Phase I, on Hosur Road, in the former Mahadevapura CMC. It is about 9 kilometers (approximately

4.7 miles) away from the southern city limits of Bangalore. In our commute from Cox Town in the northern tip of the city, we negotiated the entire north-south stretch of the city. Despite the long commute, it was a valuable way for a newcomer to familiarize oneself with the city. As the bus meandered in the early hours through the sparse city of morning walkers, milkmen, and flower vendors, the conversations on the bus focused on the progress of ongoing projects and awaiting client emails from the United States, where the previous business day had just ended. There were occasional exchanges about a new movie release or the opening of a new pub, but most of the conversation dwelled on looming deadlines or the newest version of a computer language that one had to learn soon. However, the tenor of the bus conversations was relatively relaxed compared with that within Infosys. The dark windows of the bus blurred the world outside, which passed by unobserved and, perhaps, inconsequential to work lives that were tied to the Western hemisphere.

Given my current residence in the United States, my company was of some importance to my co-commuters. For those who had stayed in the United States working at a client site, I provided the opportunity to recollect the "American way of life"; those who had never been to the United States, but wanted to, listened intently about my graduate student life in that country. They occasionally offered their own thoughts, gathered eclectically from acquaintances and the media. Having grown up in a family where some relatives lived in Europe and were married to Europeans, the conversation had a familiar ring for me. Yet there was a palpable difference.

In my childhood the word "foreign" carried a specific social status that could be displayed via the things that my relatives brought for us as gifts. "Foreign" was mostly synonymous with the "West" but sometimes could include the Soviet Union and Australia. At that time, anything "foreign," like "foreign goods," "foreign connections," "foreign vacations," "foreign domicile," and so on, was desirable but not easily accessible. To Indu and most other employees at Infosys, the West, or the United States in particular, still evokes curiosity but not the kind of desire and awe that I knew as a child. The West to them is normalized; that is, it is accessible and no longer quite "foreign." Infoscions were regularly deployed to client sites for various lengths of time ranging from a few weeks to a few years. Moreover, the regular exchange with clients via conference or video calls, emails, and client visits folded the West within their everyday lives.

Nobody denied their desire to visit the United States, "but only for work, because it is an important résumé builder" as they would usually put it. The other reason some mentioned was that they "wanted to see the land from where the software code originated." As a "résumé builder," the likelihood of a U.S. visit was also fertile ground for competition among programmers to be selected for projects that required client-site placements. However, I noticed their consistent emphasis on "visit" instead of "live." The West in this sense is a temporary geographical node to be visited but not the place to settle. Importantly, the relationship between India and the West ceases to offer a transformative experience for the urban middle class. According to most of my informants, the possibility of a "world-class life" in a "world-class city" is now available in India: "Why bother living in the U.S. away from family and friends?" The question was often gently directed toward me.

The changing orientations to the West emerge from a sense of contentment among the IT professionals who were highly conscious of their importance in contemporary India. The West, in this discourse, is another node in what Ann Stoler has called (in her analysis of colonialism), the "circuit of knowledge production" (2001: 831). The circuit of software production, which is primarily seen by the software professionals as a vital part of the current knowledge economy, evens out the long-existing geopolitical hierarchy between the West and India. The circuit draws the clients in the United States and companies like Infosys in India into a global motion. Comments like, "They need us as much as we need them" (implying American businesses) were dominant at Infosys. Or, as we have seen previously, Nandan Nilekani publicly declared such a view to Thomas Friedman in his documentary *The Other Side of Outsourcing*: "Tom, the playing field is being leveled." The West still predominates, but the advantage of India in software development lends the impression that the power differential between the two countries, as between a developed country and a developing one, are being relatively blunted and blurred.

Still, during the 2004 presidential elections in the United States the bus conversations dwelled nervously on the issue of U.S. companies outsourcing IT jobs to developing countries. Websites run by U.S. programmers who had lost their jobs to outsourcing sprouted up; T-shirts expressing agony such as "My job went to India, and all I got was this lousy T-shirt" became a common sight in areas in California with the highest concentration of IT workers. While George Bush was more inclined to continue outsourcing, John Kerry appeared to be hesitant about the business model. Indu and her friends kept themselves up-to-date with the latest exit polls to gauge the final winner and, by extension, their own future in the IT industry. As the news of George Bush's win for the second term hit Bangalore, there were signs of reprieve on the bus the next morning. Interestingly, earlier when I tried to engage the same group on the general elections in India in May that year, I received limited reaction. Neither was there much awareness among the software employees of their privileged middle-class location when they expressed their preference for a world-class lifestyle.

Sometimes the conversation focused on my work, perhaps given the remoteness of a discipline like anthropology for software programmers. Like others, Indu often asked me about my research, how it was progressing. I said, "It's going well, and Infosys *is* the place that I wanted to base my research."

She smiled conveying a quiet pride in her organization, which I also perceived in others in this context. I also mentioned to her and others I came to know about my simultaneous work with Janaagraha. Indu expressed her familiarity by saying, "Yes, I have heard about them. They are really doing great work for the city." To pursue this further, I asked, "You know Infosys and Janaagraha are working together to address the problems of Bangalore?" Indu responded, "I am not sure if it is us really, but I know Nandan Nilekani is involved. It's good work, but I just don't think we have the time for doing anything else." Some engineers in education and research (E&R) and at the Infosys Leadership Institute (ILI), aware of my work with Janaagraha, also mentioned that they were often unable to honor requests made by Janaagraha to help them with their volunteering needs. On the other side, the absence of Infosys employees volunteering was a matter of concern at Janaagraha. In several Monday Morning Meetings I attended, a few minutes would be spent in discussing how employees from the IT industry, particularly from Infosys, "could be and should be involved in the work of the city, maybe on weekends only." At this point it may seem awkward that the citizens' movement did not have a populist base in the IT industry, especially among the young software professionals. Despite their want of actual volunteering, most employees of Infosys supported Janaagraha and were particularly vocal about their mission to end corruption. Some also mentioned that if Infosys would formally organize a collaborative time with Janaagraha, they would definitely register to volunteer their time.

I posed this question, that is, whether there were plans to formalize the relation between Infosys and Janaagraha, in my conversations with upper management. The answer did not preclude such a possibility, but there was no definite plan then. While the collaboration between Infosys and Janaagraha seemed tenuous at first, the association was nonetheless deep. It particularly manifested itself in how the vocabulary in these two sites overlapped. Thus to think of the relationship between IT and the "outside" simply as a matter of direct coalition is limiting. Rather, the corporate framework based on "transparency" and "accountability" that ushers in the market as the new organizing tool, seamlessly but nonetheless troublingly surfaces in areas that were previously unconnected, such as water. As I demonstrate here, the everyday work of software development and business processes at Infosys offers a new blueprint for public governance. The blueprint is not only essential to stem "corruption" but is also an indispensable ethical charter to reorganize the relationship between the state and the citizens as "stakeholders" of the nation. Although the blueprint is certainly not exhaustive in terms of understanding neoliberal change in India, it is nonetheless a substantial intervention in recasting public governance.

In the following pages, I first offer a brief historical account of the software development industry in India, specifically highlighting how a public-sector initiative paved the way for very profitable private enterprises. The subsequent sections of the chapter discuss the workflow followed by several software development and business-processing units at Infosys. I specifically focus on how each process is designed to arrive at an unequivocal resolution. Seen another way, software development is therefore aspirational in nature because it seeks to extend the limits of possibilities. It quickly became clear to me that most IT professionals valued the unequivocal and the aspirational dimensions of their work, which they frequently characterized as "scientific" and "rational."

In turn, when I inquired, this mode of thinking informed their approach toward the nation-state, as well. There was an evident and widespread impatience with the condition of the state, especially its "corruption." As one engineer put it succinctly, "Wish we could write a software to end corruption so that nothing can work outside the loop." The connection employees at Infosys made between the neutrality of the software is comparable to the market, on one hand; on the other, it activates what I have termed here an ethico-political engagement with the state, where redemption from "corruption" is critical. I call this aspiration and desire for unambiguity the "software nation." By this I imply, first, the actual processes software development entails. Second, I wish to simultaneously emphasize the emergent blueprint being generated for the nation-state. I end the chapter by indicating some significant contradictions that surfaced in the narratives, expressly of some senior employees who had worked in other sectors before joining IT. On one hand, we notice anxiety over losing "traditional" Indian values. On the other, IT is still appreciated as a tool for national redemption. Particularly, I highlight the connection made between tradition and Hindu mythology as a way to think about India as a Hindu nation, a disquieting manifestation of Hindutva Lite.

Tracing the Software Industry

The Karnataka State Electronics Development Corporation (KEONICS), and particularly Rama Krishna Baliga, is known to have promoted Bangalore as the Silicon Valley of India and is even credited with this nomenclature as part of this initiative. KEONICS established the Electronics City in 1977. Later it extended the electronics industry to specifically include and promote the software industry. The International Technology Park, which was established in 2000, involved consultants from Singapore. The website states: "The vision that mooted the International Tech Park Bangalore (ITPB) originated from a meeting in 1992 between the then-Prime Minister of India, Mr. P. V. Narasimha Rao, and the then-Prime Minister of Singapore, Mr. Goh Chok Tong. In 1994, a consortium comprising Indian and Singaporean private enterprise was formed to spearhead Park development."[1] ITPB promoted a "Live, Work, Play" lifestyle confined within the business park. James Heitzman (2004) has discussed this history to show how the city's infrastructure morphed under public policy initiatives to privilege the flow a particular kind of global capital geared toward software production. Also, Rasmus Lema and Bjarke Hesbjerg examined the interconnections between the institutional support and sector-specific characteristics that influenced the relationship among software firms in the city (Lema and Hesbjerg 2003). The role of KEONICS, an otherwise public-sector organization, is important here as it gradually transformed to promote the private global software industry. In fact in the 1970s, most of these firms, as Heitzman has noted, were focused on import substitution. However, with the emergence of "body shopping" and later offshore outsourcing where quality software labor could be obtained at cheaper rates in India facilitated by the transnational telecom connectivity, the software industry in Bangalore emerged as a prime location for outsourcing. Texas Instruments, who was the first company to avail satellite communications, preceded Infosys and Wipro in this regard.

Joseph Grieco (1984) has explored in depth the shift from foreign to Indian computer firms between 1960 and 1980. The two-decade-long collaboration between India and international IT firms, Grieco argues, transferred the power in favor of the former. The shift was an important policy decision that predates the 1990s decision to liberalize the economy. Of particular importance here is the departure of IBM in 1978, which at that time was the largest foreign computer firm in India. Grieco writes:

> In early 1966 the government (Indian) expressed a desire that IBM allow Indian participation in the ownership and control of what at that time was one of a very few 100 percent foreign-owned subsidiaries in India. IBM responded that as a result of the independent character of its international operations, the company required highly centralized control (and, consequently, ownership) of its overseas subsidiaries, and therefore it could not share equity with Indian nationals. Indeed, IBM advised the government that the company would withdraw from India rather than share ownership. (1984: 25)

The narrative of IBM withdrawing from India was still alive among IT professionals at the time of my fieldwork. However, while Grieco notes "IBM was forced," Indian IT professionals often portrayed this as "IBM was asked to leave." The departure of IBM, most of them argued, was a valuable opening

---

1. Available at http://www.itpbangalore.com/pp_history.html (accessed December 2, 2005).

for India to create its own software industry, as one programmer put it, "on our own terms."

Rational Imaginings

Infosys is composed of eighteen buildings (more were under construction when I was doing fieldwork), each dedicated to a specified job such as management, software development, research, education, quality assurance, human resources, and so on. The glass-and-concrete buildings are numbered (rather than named). The numbers were prominently displayed close to the roof, making navigation among them easy. However, some buildings, such as the first one, which housed the management and senior officials, was referred to as the "headquarters." On my first visit I learned to refer to the site as a "campus" instead of a company. This term is borrowed from the likes of Microsoft in Silicon Valley in the United States. It was a rhetorical way of emphasizing Infosys as a space of learning. As employees learned new skills to keep up-to-date with new software trends, "campus" enhanced the pedagogic importance of the industry.

Few programmers I met majored in computer science or information technology either as an undergraduate or a graduate student. Rather, they came from various natural science disciplines, such as physics or chemistry, or from traditional engineering fields like mechanical, electrical, and so on. The Infosys recruitment policy states that there is no prerequisite to know software programming for candidates to apply for a job. However, the applicant had to be a student of natural science or engineering; students from the social sciences and humanities were not barred but discouraged from applying. I asked upper management about this particular preference. As Dr. M. P. Ravindra, then senior vice president and head of education and research at Infosys, explained to me: "In the philosophy of Infosys we do not insist on the IT background. We have specially designed programs to learn the software principles. Students who take up science or engineering have a natural bent for mathematical thinking which is rational. It becomes much easier to train them because the foundation is already there."[2]

Dr. Ravindra offered me his biography as an example of the advantages of natural sciences and engineering for IT. He holds a doctoral degree in theoretical physics from the Indian Institute of Science in Bangalore and was involved in full-time fundamental research. "But around the early eighties," he continued, "I decided to move to IT, which was very easy because I was already involved in rational thinking. But if you are from the humanities, you are thinking in terms of your feeling, not reason. I will show you a sample test, and you can see what I mean." He telephoned his assistant, requesting

---

2. Interview conducted in September 2004.

she bring in an Infosys sample test administered to college students. Typical questions might include the following:

1) Father's age is three years more than three times the son's age.
   After three years, father's age will be ten years more than twice the son's age.
   What is the father's present age?
2) Ram, Shyam, and Gumnaam are friends.
   Ram is a widower and lives alone and his sister takes care of him.
   Shyam is a bachelor and his niece cooks his food and looks after his house.
   Gumnaam is married to Gita and lives in large house in the same town.
   Gita gives the idea that all of them could stay together in the house and share monthly expenses equally.
   During their first month of living together, each person contributed Rs.25.
   At the end of the month, it was found that Rs 92 was the expense so the remaining amount was distributed equally among everyone.
   The distribution was such that everyone received a whole number of rupees.
   How much did each person receive?
   (Hint: Ram's sister, Shyam's niece, and Gumnaam's wife are the same person.)

He pointed me toward the answers and expressed, "At Infosys we focus on problem solving rather than the execution of an order, and only science and engineering students can take a test like this and do well." The rigid epistemic separation between affect and reason is not unprecedented and has a rather entrenched legacy since the independence movement. As part of the movement and to subvert Western colonial superiority, an image of India was promoted, which, as Gyan Prakash (1999) claimed, was "inseparable" from science. Prakash writes, "Standing as a metaphor for the triumph of universal reason over enchanting myths, science appears pivotal in the imagination and institution of India, a defining part of its history as a British colony and its emergence as an independent nation" (1999: 3). Scientific or rational thinking, as Ravindra put it, hence continue to dominate the imagination of the nation-state. I resituate Prakash's claim in two ways: First, there is the question of time intertwined with processes of globalization, which makes IT quite a distinct domain of this imagination; second, there is a need to recognize the changing relationship between science and India's relationship with the West, which is no longer perceived to be transformative.

As Prakash identified, the scientific "structures" in which both the co-lonial and the independent states invested, such as railroads, steel plants, mining, and so on, all drew on a vibrant pedagogy of natural sciences and engineering. These traditional disciplines provided the crucial backbone for infrastructural work during the nation-building era of India. With the advent of IT, these traditional disciplines not only receded but also ceased to be valued even in terms of national and/or individual aspirations. While the imagination of the nation-state continues to be based on science, the foundation of this imagination has shifted. The steel plant no longer stirs the imagination; a software company does. Rather than looking at this simply as a matter of scientific progress, one needs to capture the imagination of the nation-state, which is animated by software as not merely *another* technology but as a paradigm.

I posed the question to Dr. Ravindra, linking it specifically to his personal biography: Why did he consider a move to IT necessary given that he had an established career as a physicist? He told me: "There is something about IT that is absolutely fascinating. It can do almost anything; you know I mean the code. It can do so much, and the challenge is much higher than physics." Ravindra's view was echoed by others, as well. Vinay Kumar, a faculty member in the E&R department and a metallurgist by training, explained, "The traditional fields are closed and stagnant; there is nothing new happening there. But so much is happening in IT, it's a new field." Dr. J. K. Suresh, who heads the Knowledge Management Group, was originally trained as aerospace engineer at the Indian Institute of Technology, Kanpur, and at the Indian Institute of Science. He shares a comparable biography with Dr. Ravindra. The appeal of IT as a new field was evident among most employees, irrespective of whether one chose to join IT as a career following college or, like Dr. Ravindra and Dr. Suresh, was a relatively late convert. Collating what I gathered, the appeal of IT lay in its process, which I discuss in the following section.

Software Processes

Infosys works with clients from various industries, such as banking and capital markets, telecommunications, media and entertainment, insurance, retail markets, and so on. Each industry domain is placed under an Independent Business Unit (IBU), which specializes in software tasks related to its specific business needs. The IBUs are referred to as the "verticals" of the company, that is, the pillars on which the company rests. The IBUs are self-contained in that they comprise a sales and marketing team along with the software development and testing teams. At any given moment, several teams in an IBU are simultaneously working on several projects from different clients or different projects from the same client. Some teams are also dedicated to remotely maintaining live projects originally designed by the IBU. Software knowledge is not directly shared among the IBUs except in the context of a Horizontal Business Unit (HBU), where the expertise from various industry domains is combined to solve a generic business process or to enhance Finacle, Infosys's banking product. All completed software projects are collected in a centralized archive known as the Knowledge Management Division, which serves as a repository for future software work.

Most IBU work areas look similar. It is a large hall split into several small open cubicles in the center, with offices and conference rooms along the perimeter. While junior programmers worked in the cubicles, the project managers and senior members had individual offices. The space carries a distinct patter of computer keyboards, broken occasionally by a human whisper now and then when neighboring colleagues may peek over their cardboard separation to exchange a few words. Overall, it is a disciplined space, quite distinct from my experience at the Water Board, which, being a public space, was noisy and constantly bustling with people.

Like other software companies, Infosys draws upon the standards set by Carnegie Mellon's Software Engineering Institute (SEI). SEI focuses on "process maturity" where companies are evaluated on what is known as the Capability Maturity Model (CMM). The CMM ranking ranges from 1 to 5, named as "initial," "repeatable," "defined," "managed," and "optimizing," indicating increasing maturity, respectively (Pressman 2009). By the time I was doing fieldwork in 2002, Infosys had achieved the final stage of maturity, CMM level 5. Level 5 indicates that along with detailing and managing software development, Infosys was also engaged in testing innovating ideas and technologies. This certification was crucial for Infosys to retain existing clients and attract new ones. "It proves," as one senior project leader told me, "that we too are capable of meeting higher standards despite the fact we are in the land of corruption."

The CMM in turn is based on a variety of software process models used to design the flow of the codes. Roger Pressman (2009) argues that coding is a method of dealing with and "bringing order" to an otherwise chaotic situation. The software industry usually follows a number of available models for processing the work. One often used at Infosys was called the "Linear Sequential Model." This model has four distinct steps: requirement analysis, design, coding, and testing. The first step involves an engaged discussion with the client to understand his or her business processes and how software can be integrated to streamline the workflow, remove specific intrasystem barriers, or enhance the current state of functioning. This in fact is the most intense part of the project; the emphasis lies in arriving at an understanding of the business needs and then figuring out the ways software can be developed to address those needs as closely as possible. Any gap between the client's expectations and the software design can escalate to a "business risk," something I discuss later.

Once the requirements are decided and settled, Infosys and the client sign a document of agreement. This document leads to the design phase. The design phase focuses on four areas—data structure, software architecture, interface representations, and procedural details, specifically algorithmic. This process translates and represents the business requirements as a software design. Next, the coding makes this design readable by the computer, where the execution of the software is completed. The final phase is known as testing, where the program is tested to assess its integration with the business logic and workflow. This is also an anxious phase for the project team since the software is also now made available to the client for evaluation. Almost all projects are subject to another round of changes, which are usually minor. However, in certain instances a project is significantly overhauled, either because there is a gap in the requirement analysis or the needs of the client have changed dramatically. Once the testing is completed, the software "goes live" and the client starts using the software in everyday business, the project enters the "support" phase. Support involves regular maintenance of the software and also fixing errors that may interfere with its regular functioning. The same team may continue to work during the support phase, which, depending on the contract, can range from a few months to a few years, or a new team may be formed for the purpose.

Another model of software development, which Infosys occasionally followed, was called the Prototype Model. The Prototype Model is adapted when the client only has an amorphous idea of the kind of work they want the software to achieve and was also unsure about the kind of human-machine interaction they wanted for his or her customers. The team in this case gathered the minimal requirements available and swiftly put together an application that worked as a functioning software to test if it could or did address the business needs. This sample software is then modified on the enhanced business understanding of the client and the recommendation of Infosys.

Though software development, like most other industries, is necessarily team-based work, there is nonetheless an overwhelming focus on the individual in the IT industry. By "overwhelming," I suggest that while corporate promotions are not new, the IT industry works with another kind of "process maturity" model, similar to the CMM I discuss above. The People-CMM framework was also developed by SEI and was first released by in 1995. It was described as

> a framework that helps organizations successfully address their critical people issues. Based on the best current practices in fields such as human resources, knowledge management, and organizational development, the People CMM guides organizations in improving their processes for managing and developing their workforces. The People CMM helps organizations characterize the maturity of their

workforce practices, establish a program of continuous workforce development, set priorities for improvement actions, integrate workforce development with process improvement, and establish a culture of excellence. (Curtis, Hefley, and Miller 2009: vii)[3]

The first step, which in Infosys was called the "induction" phase, involved a relatively unrestricted program where new employees were introduced to various programming languages and concepts and were only required to practice their skills in a simulated environment. This phase also familiarized them with the work culture and values of Infosys. However, increasingly, there was a calculated focus on managing the available human resource, such as identifying and cultivating individual competencies and then deploying them to specific projects where these competencies can be best optimized.

In the Foucauldian sense, the People-CMM framework is a digital mode of governmentality. It tracks, classifies, and deploys the existing human resources at Infosys for maximum economic productivity. Such corporate hierarchies are not unprecedented. However, since the 1990s, especially with the advent of IT, there has been an overwhelming focus on how one can best align individual capacities and competencies with the business requirements. This alignment is also unique in the sense that this is computed quantitatively. Governance, that is, corporate governance in Infosys, was thus an alignment and the maximization of individual and enterprise productivity. Unlike former corporate practices, Infosys actively worked to identify and harness individual capacities to enhance its business. In the following sections, I offer instances of the various stages that combine both IT and people processes at Infosys to maximize productivity.

Induction Program: UNIX Day 2

I had a quick lunch and headed for Building 12, which houses the E&R and knowledge management (KM) wings of Infosys. Unlike the other buildings, which have an austere rectangular shape, the E&R building is a rotunda. The Cartesian logic—the E&R building in the center and other buildings placed along its axes—as many told me, indicates the central role of knowledge in Infosys. The E&R and the KM groups are responsible for accumulating, archiving, circulating, and charting the software knowledge that is generated as projects are completed throughout the organization.

I signed my name at the security desk, which was a standard practice I had to follow to enter any building. The security officer checked my "Visitor" identity card that allowed me access to the entire Infosys campus, or "Infy

---

3. Available at https://www.sei.cmu.edu/reports/09tr003.pdf (accessed September 22, 2010).

City" as the card noted. He nodded his head and picked up the phone to call Vinay Kumar. I was there to attend Kumar's introductory class on UNIX for the new campus recruits who were currently undergoing the induction program. I attended several of these software introductory programs. Since most of them follow the same pedagogic structure, I am presenting this as a typical case. Dr. Ravindra had mentioned me to Vinay, so on the phone he asked me to go to the auditorium where the class would begin in the next ten minutes. Before entering the dark auditorium, I looked up at the rotunda's glass ceiling through which the afternoon sun poured in to obscure the concrete-and-mortar structure.

Distinct from the professional milieu that I had encountered thus far at Infosys, the atmosphere here replicated a classroom. Young men and women, who definitely looked like college students with their carefree attitudes and frequent giggles, filled the auditorium; they sat spread out in small groups of five or six. I introduced myself to Vinay as he was setting up his laptop with the projector for the PowerPoint presentation. Vinay mentioned that this was the second session of the required class on UNIX for all new campus recruits.[4]

Vinay started by assigning the class to write a file name as the first step in a software code. The projector also showed a few lines of code that appeared in rapid sequence. He wanted to know how many in the room were familiar with the set of commands UNIX uses. Some raised their hands immediately, and some did after looking around for some support. Vinay directed his gaze directly to the ones whose hands were not raised and said, "I will give you practice assignments." Vinay's position at the podium, like a teacher, commanded respect: the class referred to him as "sir" while answering or asking questions, which is the usual norm in Indian academia. However, it is relatively unusual in the corporate culture of Infosys, where, with the exception of Narayan Murthy, who is referred to as either NRN or Mr. Murthy, even Nandan Nilekani was referred to by his first name.

Earlier Nandan Nilekani had mentioned this informal attitude of the company to me with pride. To him, it is a form of leveler in a society where showing deference to the powerful and the elderly is a long-established hierarchy: "Being a company that is comprised of relatively younger people, we need to break the old habits, which really do not have a rational basis and it creates all kinds of irrational hierarchies." He continued, "My doors are always open, and if anybody wants to talk to me about some idea, they can

---

4. UNIX is a popular multiuser, multitasking operating system developed at Bell Labs in the early 1970s. Created by just a handful of programmers, UNIX was designed to be a small, flexible system used exclusively by programmers. UNIX was one of the first operating systems to be written in a high-level programming language, namely, C. This meant that it could be installed on virtually any computer for which a C compiler existed.

just give a call and walk in." Other Infoscions I spoke to later ratified this. The website described working at Infosys as follows: "We would like to describe our people and our work place in simple terms. But it isn't easy when what we are trying to describe is a certain feeling of *joie de vivre*; a feeling of energy and vitality, of freshness, of a place where people work in a campus like facility and culture, are unafraid to voice new ideas, of a place where there is minimal hierarchy."[5]

I asked if this wasn't also an effort to emulate practices in the United States. While my informants acknowledged that, they were also quick to negate it by mentioning that it "does not have to be of the U.S. forever," indicating once again a changing relationship with the West. However, the aspiration to eradicate hierarchies of power, age, and so on, is not evident throughout Infosys, as is evident in the induction program I discussed above. The aspiration was compartmentalized and for a specific reason: Infosys's investment in being a world-class company was also tied to pedagogy. Pedagogy in the traditional Indian education system is based on an authoritative relationship between the teacher and the student. It is different from merely teaching or coding software; pedagogy indicates a method and a passion for knowledge. Also, since Infosys describes itself as a campus to simulate a university or college impression, the pedagogical norms of interaction were prevalent and valued.

In this vein, space and time were rigidly structured, as well. As a human resources executive clarified to me, "Breakfast and lunch are served only during specific hours of the day; the gym and the swimming pool can be accessed after 5 P.M., only at the end of the workday. But the Coffee Day is open throughout the day."[6] Every building had a clocking-in device at the entrance foyer where Infosys employees swiped their ID cards four times daily—in the morning before beginning their work, in the afternoon on the way to lunch and back, and finally when leaving in the evening. The magnetic strip would record their hours of work and breaks taken in between. These are usual corporate practices of controlling labor in relation to capitalistic logic. In analyzing transitions to the industrial notion of time, E. P. Thompson (1967) has called this the "time-thrift." He writes, "In mature capitalist society all time must be consumed, marketed and put to *use*. It is offensive for the labor force merely to pass the time" (1967: 90–91).

Yet what is specific about the nature of spatiotemporal organization of IT? How is pedagogy and discipline intertwined in a context to augment profit?

---

5. Available at https://careers.infosys.com/infyrms/infycareers/careers/workatinfosys.asp (accessed March 13, 2006).

6. The café chain in urban India is formally known as Café Coffee Day and informally referred to by its acronym, CCD. Infosys had several CCD outlets spread throughout the campus.

The precise organization of space and time at Infosys not only signified a modern industrial discipline; it also marked a separation. It was a conscious dissociation from the generic Indian perception of time and space as fluid and abundant. The sardonic twist of the maxim, "Indian Standard Time," or IST, as "Indian Stretchable Time" I mentioned earlier is an instance where time is perceived to be indefinitely elastic. It flows interminably defying the limits of hours, minutes, and seconds. The use of space is similar; one kind of space spills over into another where boundaries are inconsequential. The government offices, for instance, as I mentioned earlier, double as an active space for both work and socialization. An allegiance to a restricted spatiotemporal notion reflected the modern work ethic of Infosys as well as its reliability and accountability to stakeholders. Such spatiotemporal practices were also visible in Janaagraha. We may recall the rule where anybody arriving late for the weekly Monday Morning Meeting was required to pay a late fine of five rupees. Software coding as a practice aligns with Thompson's notion of time-thrift while also recasting time as a valuable resource, which can no longer be squandered.

Let us return to Vinay's classroom. After writing a few lines of code on the whiteboard, Vinay introduced the "building blocks" for writing software codes to the class:

> To write the program, build the logic first; syntax is the important feature of the coding language. . . . But can one write a bug-free software code? Keeping error to the minimum and overriding default behavior are the keys of a good software program. . . . There are different levels of intelligence in the software program and so managing the history is important and you can use the "search" tool to do that . . . "search" is a very important tool.

The class listened intently to Vinay, raising questions now and then to clarify the lines of codes that appeared on the projected screen. Occasionally some would get distracted and Kumar had to get their attention back. While most of the class concentrated on writing the codes and following the instructions, some did not; they either hovered over their colleagues' notebooks or watched the screen in front. Vinay, noticing this, reminded the class that "programming can only be learnt hands-on, doing it oneself and never by watching another person do it."

I recalled my days of learning software coding, and the inclination to mentally pursue the steps without actually writing down the code since it looked so logical. The logic of the code, also known as source code, is a series of statements tied to a rigid syntax. In other words, the logic is embedded in the syntax and flows through it. Thus the real challenge of software coding lies in seamlessly including the logic in a structured manner. The source code is written in a human-readable computer-programming language. It is a collection of files specific to a piece of software. The physical scripting of the code along with the mental work is encouraged in the industry as a way to master the structure. Further, since the same source code can be and is used by more than one project or different programs in the same project, the utility value of the code is high for companies like Infosys.

The structure of the source code is in turn embedded in the *loop*. The loop is the overarching logic of the code, and for a software code to be error-free or "bug-free," it will have to complete its given loop to move on to the next one. The source code is either converted into an executable file by a compiler for a particular computer architecture or executed on the fly from the human-readable form with the aid of an interpreter. The compiler generates a list of errors, which then have to be removed; the process is usually known as debugging. While bug-free code is almost mythical, the aim is to minimize the errors for a loop to be concluded.

By the time the class ended after two hours, I had collected a list of acronyms, some of which I could expand from my own knowledge of UNIX, others I saved for Vinay. ASCII stands for American Standard Code for Information Interchange, BS for backspace, CAN for cancel, CR for carriage return, and so on. Though acronyms and abbreviations organize software development, they are not exclusive to the IT industry. However, being familiar with the IT industry is to know, recognize, and employ abbreviations swiftly that matter in everyday work. Though there are different forms of abbreviations, the nomenclature of acronyms, "alphabetism" and "initialism" are most common.

My interest in abbreviations is in what they signify for those who use them. That is, IT professionals often saw abbreviations as a timesaving device. I argue that while it compresses time, it also compresses the social into an instantaneous and neat expression that is readily exchanged and understood. The software code abbreviates human transactions within the limits of a predetermined loop. Moreover, both the code and the acronym coalesce the social, the political, and the economic in one single and specified performance. Then how does the code organize social life, its praxis and imagination? Can we think of the code as a tool to reimagine the nation-state? What does the code offer in terms of this imagination that is not only new but also in some ways cathartic?

Memory and Debris

I met Vinay again later that week. He explained that he was primarily responsible for the training program of E&R, such as the UNIX class. He told me he conducts training at various levels, "entry, advanced, middle level, and also upon request." A phone call interrupted our conversation; Vinay excused himself to answer the phone, which was "for an important request that has to be scheduled urgently because this team is leaving for the client site in the U.S." "Training," Vinay stated, is a "vital organ of Infosys because of the changing nature of the knowledge base; there are new programs coming out every day and we have to keep everybody updated." I was curious about the nature of the change. Does it destabilize the knowledge base completely or is it incremental?

Vinay explained:

> The changes are built into the fundamental concepts of software, so it is easy to pick up. But this change is very necessary; it is this change that makes IT different from the other industries, it makes it more dynamic. But again remember, dynamism does not mean revolutionary. It is not like the Internet revolution that changed everything for us. Software change is not difficult to cope with once you have the basics of coding right.

Vinay compared software programming with firefighting. "It is like taking on the fire head-on because you know you have to survive, it is not an option. You are trained but you also learn on the job. IT is also about self-learning. It's either do-or-die; it's tough." The demand to remain up-to-date with the latest developments of software technology was evident in all my conversations with Infoscions. Ignorance of a new software language triggers a sense of anxiety that one may be excluded from more challenging new projects.

In a self-test, administered at the beginning of a "time management" workshop I attended, junior and midlevel programmers reflected high levels of anxiety regarding their pace to learn new programming languages. In the bookstores downtown, the software-language books were religiously placed near the entrance of the floor that sold "Computer Books." The knowledge of a new language and the ability to write programs using it is not merely a "résumé builder." On several occasions I witnessed, only those deemed the best programmers were called on to resolve a specific problem that was stalling a project. The best programmer, as I gathered, was one who was not only adept at coding and solving problems but also a quick and dedicated learner always conversant with the latest programs. The pedagogical commitment of Infosys bolsters and creates an ambience of learning where, as one Infoscion put it to me, "one cannot help but feel like I am back in college."

What does one do with languages that are no longer viable and usable? One can never be certain when one programming language will be rendered useless by the arrival of a new one or when an older language may resurface with a vengeance. "Like Infosys, programming languages too have a market of their own," Indu explained to me once. "So our challenge is to figure out which programs will be best suited to build an application that the client wants. We have to do our best because Infosys has signed a contract with the client for a great deal of money. If the client is not satisfied, they will go to another company, and there are plenty of them in Bangalore. And, well, our compensation packages and our promotions are also tied to how well we can do the job."

"So what do you do with the programming languages that are no longer useful?" I was curious. "We just let it go." She added, "We have to." What seems like a cavalier attitude toward knowledge in effect constitutes the very ethos of the industry. Like commodities, software creates its own debris. This debris is not necessarily seen as an impediment. Rather, it enables innovation and deters stagnation, both of which are appreciated as hallmarks of a dynamic field like IT versus the traditional science and engineering disciplines. As some employees put it, "The only way we can know if IT is progressing is in the ability of the program to work on problems better and faster." The relationship between progress and stagnation, innovation and debris led me to the issue of computer memory in particular and, related to this, the overall notion of the past in the IT industry as a whole. These questions are relevant, and I discuss them in Chapter 4, because they raise the issue of the past, which was often perceived by many of my informants as burdensome and obsolete.

Standardization and Differentiation

As I emphasized earlier, the IT industry in India is primarily dependent on the software export to businesses primarily in the United States and Western Europe. This makes the industry heavily reliant on the politics and economies of countries where they have no control. In this context, I was curious about the domestic facet of the industry—that is, what about clients within the country? At the time of my fieldwork, the number of domestic clients was minimal and was not even an important conversation in Infosys. Around the time of this research, however, NASSCOM reported that the domestic IT market grew 22.4 percent in 2007.[7] E-governance is a central area where growth is expected along with larger Indian firms outsourcing the IT jobs "comparable to the global deals in size."[8] NASSCOM, however, notes several constraints that are hindering the deployment of IT in India. For example, Indian businesses invest only 1.5 percent of their revenue, compared with those in the United States and United Kingdom, which invest 5.5 percent, indicating the slow pace of government initiatives in e-governance. The

7. Available at http://www.nasscom.in/upload/57956/Domestic_IT_Market_Trends.pdf (accessed December 14, 2010).
8. Ibid.

business-to-consumer model via the Internet is comparatively low, which is also cited as another thing that distinguishes the minor presence of the domestic IT business.[9]

In the edited volume *From Underdogs to Tigers: The Rise and Growth of the Software Industry in Brazil, China, India, Ireland, and Israel*, the section on India identifies the difference between the domestic market and the foreign market, the latter being the main focus (Arora and Gambardella 2005). The domestic market is geared toward the sale of software products. Infosys's Finacle, a banking software product, is one such example that was sold to nationalized banks mainly within the country. However, though Infosys lists several products on its website, Finacle is perhaps the only one we can consider on a par with those of, say, Microsoft. The rest are basically methods to augment software performance, a task undertaken by SETLabs, which I discuss below. However, Arora and Gambardella (2005) also argue that the industry provides a much wider range of services for the domestic clients that in ways are also more challenging. They cite the example of digitization of the Reservation System for Railways completed by CMC, a subsidiary of Tata Consultancy Services and a long-established (1975) public-sector firm, which involved a prolonged commitment to every phase of software development and maintenance.

Minor innovations aside, the software industry overwhelmingly follows standardized methods of work. The relationship between standardization and innovation is delicate. Standardization involves a set of rules that needs to be meticulously followed if one has to compete in the market, but it can also potentially undermine innovation. As Xiang Biao (2007) has noted, the global IT trade seems to have reached a stable but troubling division of labor. While companies such as Infosys in developing countries like India ardently observe standardized modes of work, which can be replicated, companies in the West focus more on advancing the technology.

Considering this clearly hierarchical relationship, the (older) tagline of Infosys, "Powered by intellect, driven by values," is rather incongruous. Is software programming intellectual and/or innovative? When I mentioned this imbalance to Nandan Nilekani, he argued, "Writing software programs and developing programming languages are both challenging but in different ways. Here we work with the client to solve his business needs and not just develop software for them. We try to locate and solve their problems and also help them streamline their businesses by giving them constructive suggestions. It is an integrated approach and not just software work we look at."

---

9. The domestic market has grown since then. For 2015 NASSCOM projects the domestic IT-BPM (business process management) sector to grow over 14 percent and earn revenues of around $50 billion dollars, which is close to half the export revenue. Available at http://www .nasscom.in/domestic-itbpo (accessed April 17, 2015).

Likewise, a related issue that most Indian IT companies confront in the outsourcing business is how to distinguish themselves from other fellow companies. For instance, Wipro, another prominent company, promotes itself as follows:

> Wipro Technologies is a global services provider delivering technology-driven business solutions that meet the strategic objectives of our clients. Wipro has 40+ "Centers of Excellence" that create solutions around specific needs of industries. Wipro delivers unmatched business value to customers through a combination of process excellence, quality frameworks and service delivery innovation.[10]

Tata Consultancy Services (TCS), by far the largest IT company, describes its mission in the following words:

> To help customers achieve their business objectives by providing innovative, best-in-class consulting, IT solutions and services. Make it a joy for all stakeholders to work with us.[11]

Infosys's is similar:

> Infosys provides end-to-end business solutions that leverage technology. We provide solutions for a dynamic environment where business and technology strategies converge. Our approach focuses on new ways of business combining IT innovation and adoption while also leveraging an organization's current IT assets. We work with large global corporations and new generation technology companies—to build new products or services and to implement prudent business and technology strategies in today's dynamic digital environment.[12]

While the IT industry in India appears homogeneous and is collectively referred to as "Indian IT companies," the internal competition among the companies fractures this identity. Infosys invests heavily on distinguishing itself as a unique "Indian" company. Infosys particularly highlights its research dimension in this context to assure clients that their work in software development draws on a robust knowledge base of IT. However, it is not difficult to see that this rhetoric of differentiation is fairly common. Differentiation to Infosys is, nonetheless, a constructive corporate rhetoric. The vocabulary is syntactically dispersed in and through various modes of interaction: excellence,

---

10. Available at http://www.wipro.com/aboutus/whoweare.htm (accessed January 18, 2006).
11. Available at http://www.tcs.com/investors/BusinessOverview/Introduction.aspx.
12. Available at http://www.infosys.com/about/default.asp.

business solutions, value, strategy, innovation, and so on. Awards are another form of differentiation. Infosys listed the awards they received in 2004 on their website:

- Infosys has won the Golden Peacock National Quality Award for 2004.
- SAFA (South Asian Federation of Accountants) Best Presented Accounts Award 2003 in the Communication and Information Technology Sector based on the evaluation of the Annual Report of the company.
- Award from Global Finance magazine as the Best company in the Computer Software Sector in Asia.
- Infosys Technologies has won the prestigious Global MAKE (Most Admired Knowledge Enterprises) award, for the year 2004. Infosys won the award for the second time in a row, and remains the only Indian company to have ever been named a Global Most Admired Knowledge Enterprise.
- Nandan Nilekani rated among the world's most respected business leaders in *FT-PwC* survey.[13]

Awards, as several managers told me, are obviously considered a better marker of differentiation because they address specific areas of the business and are also conferred in an open competition organized by independent agencies. But awards are also replicable; different agencies offer different companies similar awards. My intention here is not to seek the basis of differentiation but the *necessity* of differentiation in an industry that in the end relies on standardized methods of organizing work and human resources.

The Solutions and Software Engineering & Technology Laboratories (SETLabs), the applied research unit, was established as a marker of such supposed differentiation. Srinivas Thonse, a senior architect of SETLabs, is acutely aware of his task to "be able to make Infosys stand out in the IT business." SETLabs, he explained, "focuses on efficiency, emerging areas of IT and the market, and prepares them (programmers) with a readiness to learn new technologies." In this context, he introduced me to a process developed by SETLabs named InFlux:

InFlux™ is the Infosys methodology for defining effective IT solutions for today's enterprise initiatives. The framework facilitates the analysis and optimization of business processes to determine the impact of the new initiatives and definition of IT solutions for such business initiatives. An integrated framework for business modeling is crucial,

13. Available at http://www.infosys.com/about/all_awards.asp (accessed January 18, 2006).

as it becomes imperative for organizations to constantly iterate and align their business and technology strategies. InFlux™ thus forms the basis for Infosys' broad range of consulting and IT services to realize such intertwined strategies and technology driven transformation initiatives.[14]

In responding to my question about standardization, he said, "You are right. We do work with standardized languages and programs, but what seems standardized also has gaps. We are not just working on software applications but working on a business solution for the client. The clients initially seem to know what they want, because they are talking about a problem. The gaps still remain because they are not always clear on the requirements. Part of our job is to identify the gaps." The "gap" that Thonse mentions are potential impediments toward building what is called a "robust application" that has minimal errors. At the end of the project, more often than not, the gap is not a technical lapse but a difference in what the client imagined the application would achieve and what the final product yielded. One may recall this was the "Requirement Analysis" problem I mentioned toward the beginning of the chapter. Thonse mentioned another interesting fact: "Clients, most of who are not familiar with the nitty-gritty of IT, think that it can do anything, and in fact it can do quite a bit but not everything."

Bikramjit Maitra, who headed the talent development initiative at Infosys, also reiterated Thonse's observation. He offered me an actual situation from some years back when a project he was managing was approaching the deadline:

As the deadline neared we invited the client for a test run before going live. It is customary for all projects at Infosys to do so; it's part of the contract. So the testing begins, but the client is not happy. They tell me that this is not what they wanted because it is leaving out some of the vital business processes. There was little to do at that time, but the client pressed for changes that were basic. I told them, imagine you built a house on a plot of land but then you find out that at the end that it is encroaching upon your neighbors land by, say, two meters. It's not your fault, but, say, an oversight. Would you be able to move your house? It seems only then they understood that not everything can be done as easily with IT as they initially thought.

Most clients of Infosys and in my own experience as a software programmer as well, imagine IT as a panacea. This imagination reveals an expectation on the part of clients: one frequently heard: "I'm sure there must be some way

---

14. Available at http://www.infosys.com/gdm/influx.asp (accessed March 13, 2007).

to do this." Part of it, I contend, is fueled by the invisibility of the actual task performed by the software applications that remain buried in chips, bits, and bytes. In the case of IT where everything seems to work by virtue of a click of the mouse, the ease of the click, in turn, raise expectations. In this sense, the field of possibility expands indefinitely for the client, without much heed to the limits of the available programming languages.

"Things have gotten better over the years" Maitra concluded, "but it has not disappeared totally. Clients still like to believe that IT can do magic for them, and, given the amount of money they invest, they think IT *should* do all they want!" InFlux, thus also dubbed "Business-IT Alignment," operates in three consecutive phases, research, pilot project, and finally deployment. Thonse and his team do not work on client projects but start with a problem that has been reported by one of the development units at Infosys at some stage of the project. Thonse elaborated: "The problem can be technical, business, or a mixture of both. Our job is to look at it holistically not separately because they cannot be separated. A business requires IT to help the business and not something else."

SETLabs prepares business cases "every year for every research" they undertake. These cases, Thonse further explained, "are then analyzed to filter the central problems faced by software development when aligned with the business need it is expected to address. We locate the problem and solve it as a set of client benefits, the market, the timeline involved, and the overall effect. Then it is reviewed and sanctioned by SETLabs." He pointed to a section in the handout on InFlux he had given me as a reference and read it aloud himself:

> INFLUX prescribes a business process-centric approach to enable all stakeholders to understand the business problem space uniformly and to visualize how IT changes will help optimize the business processes. The IT changes are seamlessly translated into system requirements in order to achieve the targeted business solutions. . . . INFLUX is a rigorous methodology that leads to a reduction in defects during requirement discovery and analysis and leads to a significant saving of time of time and effort to ensure alignment of the IT solutions with the business objectives.[15]

One of Thonse's colleagues, Ramanujam, who was also present during our meeting, smiled. Once Thonse finished reading the paragraph, he declared, "Actionized." "That sounds like a film set," I said, and he laughed aloud and agreed. "Yes, you got it. It is like making a movie. Every good director works with the problems he faced in making the earlier movies. He learns from them and corrects them; it is never done." His filmic analogy unexpectedly lifted

---

15. "InFlux: Business-IT Alignment."

the curtain on a different world, and for some time we exchanged titles of our favorite films, focusing particularly on Bollywood. However, we quickly returned to honor the silence that prevailed among the row of cubicles where the SETLabs team was at work. All one could hear again was the patter of the keyboard occasionally broken by voices from neighboring cubicles or a lone phone conversation. He qualified his "movie set" by adding, "But it takes about three to four months to solve a problem and to develop a concept. We all have IT knowledge, but it is not always easy to use it creatively to meet the clients' needs. This is where the creativity of software lies and we try to increase it as much as we can."

As Ramanujam was speaking, the following section of the handout on InFlux caught my attention: "Infosys Predictability":

Infosys Predictability ensures that you sleep well at night
- Predictability of transparency: driven by organizational values
- Predictability of delivery: driven by processes, tools and methodologies
- Predictability of commitment: driven by people and relationships
- By combining tools, methodologies, processes, knowledge management, business insights, we deliver not just solutions, but also peace of mind[16]

With "predictability" and the concurrent "sleep well," "transparency," "commitment," and "peace of mind," the rhetoric shifted from a technical to an affective mode. The addition of affect gave what Thonse and Ramanujam called "a human face to software." "Predictability is handled mostly by the Quality Assurance [QA] division. We are just into solving problems; we do not predict problems," they told me. They advised me that in order to know more about "sleeping well," I had to talk to Satyendra Kumar, who headed QA. Thonse handed me a piece of paper, saying, "Here is his number."

Custodian of Values

I called Satyendra Kumar and mentioned my interest in understanding "predictability" in software development. We scheduled an appointment for the next day. The QA department was located farther away than the buildings I was visiting till then. Satyendra Kumar, usually referred to as Kumar, greeted me in Bengali, with the customary, "Kemon Achhen?" (How are you?) "A Bengali, I see from your last name. I know Bengalis well; my landlord was

---

16. "InFlux: Business-IT Alignment."

Bengali. You guys sing the Tagore [Rabindranath Tagore] songs really well, can you?" I nodded in affirmation.

Kumar started by telling me the following: "Prediction is about risks. Being able to address risks is linked to the quality of the software one writes. One can encounter many kinds of risks in software development arising from different situations. Say there is a skill issue, maybe the client is not forthcoming in time with their expectation. There could be a network problem, and not to forget the attrition in this industry. In 2004 ten projects were in the high-risk category." Kumar noticed my inquiring look as soon as he mentioned "high risk." "So there is a classification of risks?" I asked. He responded, "Yes, there is what we call low-risk and high-risk projects, but keep in mind, every project the moment it starts, starts with some risk. Say, in 2004 we had ten high-risk projects, but not all projects will start at high risk. They may start low and then elevate. The elevation is crucial for us because that's where it escalates and can get out of hand. We use a methodology called the dashboard for this."

From what Kumar explicated, software projects have built-in risks. "Is there anything specific about software risks, because every work has its own kind of risks?" I wanted to know. "Software is like any other work; they all have risks. But here we are addressing the risk and taking care of it in a focused manner; we are ensuring the quality not just of the finished product but also the process of working."

S. Lakshmi, one of Kumar's colleagues, who I spoke to soon after, further elaborated on the idea of managing risk. I present an excerpt from our conversation to offer an idea of how I engaged Lakshmi to explain the idea of risk to an outsider to the industry like me:

LAKSHMI: IT itself is quite new, to take it in the context of manufacturing and things like that . . . So quality in IT is . . . in fact may be about ten or twelve years old so not . . .

SDG: I'm sorry if you don't mind my interrupting when you say quality is kind of ten years old; I mean before that the processes that were in place—possibly there was a check on quality—some way but it was not standardized? Is that what you mean?

LAKSHMI: Exactly. It was more person-to-person than anything else. It was more like "I feel this is very important" and . . . See, quality in that sense you have to ensure it. One is the verification which is more from the testing kind, that is of quality . . . and ours is more in the assurance factor. OK so we have to bring in processes to see that a certain job—I mean—the whole process is followed in the project life cycle somewhat standardized.

Lakshmi, while being aware of the newness of IT, was also aware of its fragility. Kumar too mentioned that in such a situation of double bind, while one is striving to standardize programming, arriving at standardization itself is a challenge given the changing nature of the technology. "A young industry is more prone to disruptive technology, thinking that disruption will be evolutionary. But change should be incremental, which gives a better sense of security, but it is also a survival issue for an industry which is very competitive." The usual tension between standardization and innovation resurfaces as it has been in most of my conversations at Infosys. As Lakshmi argues, QA in IT over the years has dissociated itself from an amorphous collection of individual opinions to a systematized process that is designed to forecast, isolate, and manage perceived or actual risks.

Both Kumar and Lakshmi individually stressed the fact that though their work does not directly earn revenue for the company, it is nonetheless crucial. Lakshmi thought of their work as "some sort of a conscience for the whole of the organization when it comes to quality," while Kumar stated he is the "custodian" entrusted with the job of "gatekeeping." For the first time, at Infosys, I did not encounter the habitual phrase, "At the end of the day there is money to be made." This was often used to explain the usual pressures of software work, the deadlines and datelines that organized individual lives, aspirations, achievements, and also failures. However, Kumar's and Lakshmi's open avowals of their nonrevenue status in the company encouraged my thoughts in a different direction. Does the link between revenue and work confer legitimacy in the industry? Is this link the *only* legitimate way to belong, thrive, and succeed in the industry? Does the absence of direct revenue generate anxiety? Besides, Lakshmi's use of "conscience" and Kumar's "gatekeeping" alluded to something beyond revenue or money.

"Sure, your work may not earn revenue, but it does so indirectly. If I understand you, your job is more fundamental; it assures the company of its revenue." At this point Kumar rose from his plush leather chair and proceeded to the whiteboard on the wall in front of us and drew a basic visual to "help me understand."

Next, he explained the diagram to me, pointing out its various parts. "This is very basic, just to give you an idea of where we come in," he said. "We have teams dedicated to monitor different projects in terms of their commitment to clients, if they are adhering to the norms. Nonconfirmation leads to escalation of the risk and immediately shows up on our system." The ethical underpinning of his explication is obviously clear. He continued, "When I know that a project will finally go into critical risk, I immediately brief Nandan [Nilekani] before the client does. But first we try for an immediate mitigation. If that fails, the question is do we inform the client? You know that one of the values of Infosys is transparency, so we are always up front with the client." "This could be a difficult situation for the company," I said. "Yes, it is difficult," he responded, "but we believe that we owe it to the client. They have believed in us, and we have to keep their trust."

As Ulrich Beck has argued, if "the risks of modernization are scientized, their latency is eliminated" and hence the related question of trust cannot be overlooked (Beck 1992: 154). Who do or *can* we trust or can we trust anyone at all? Or do we replace social trust with trust in abstract and /or panoptical processes that work independently of people? These are some of the questions that I want to raise. Drawing on Anthony Giddens, "The abstract systems of modernity create large areas of relative security for the continuance of day-to-day life. Thinking in terms of risk certainly has its unsettling aspects . . . but it is also a means of colonizing the future . . . on the level of everyday practice as well as philosophical interpretation, nothing can be taken for granted" (1991: 113–114). IT leans more toward the future than the present, while the past is carefully rinsed. The everyday vocabulary, such as deadlines, résumé builders, career prospects, potential clients, expected hires, projected revenue for the fiscal year, and so on, indicates the imagination of a certain kind of future.

The abstract system Giddens analyses or specifically the dashboard that I mention above is a device to ensure a "perfect" future. It dissociates the onus of identifying risks from the individual and places in within the neutrality of an automated system. Beck's convergence of the "distribution of risk" and "individualization theorem" in modern societies is useful here. Yet, as Kumar mentioned, project managers are held accountable when their projects are labeled "high risk." "First we let the IBU head know that their project is being escalated to the critical risk. IBU heads often appeal to us and we ask them to show us the documentation of the minutes of the meetings they held with clients. For example, if they can show that the client had agreed to more time at an earlier meeting, then we may consider reducing the level of risk. We understand that not everything is in your hands, but a lot is. High risks do not indicate things have gone wrong later but that it started on a wrong note."

Ajantha Bose, then a junior colleague of Kumar, described her understanding of risk in a similar manner: "The project managers and the project leaders are the first to realize when their projects go to high risk. The 'process adherence indicators' already signal this. So it is important that one does a very detailed planning of the tasks, resources, and the risks in the very beginning. It is not just about delivering a software but also managing the collaboration among the various stakeholders." "Does this affect their career prospects?" I wanted to know. "It certainly does because when you sign off on a contract, you are taking the responsibility of completing it in time. It is a value that you should be committed to. It is not like the government, that you say something and you do something else." The repeated reference to the normative ran very deep in Infosys. The pride of being at the forefront of something that is not only technologically leading edge but also ethically empowering was explicit in the majority of my conversations.

"My Worry Kit"

The excessive allocation of resources and the lure of IT have generated some valid concerns about other fundamental science and engineering disciplines being consequently disregarded in the economy. More importantly, this concern relates to the nature of the Indian IT industry that focuses mostly on routine software development, rather than on the innovative and more critical task of designing products. As Narendar Pani expressed in one of our conversations, "IT companies unlike, say, Microsoft are focused more on software development than on designing their own products, what good will this bring for the long term future of the country?"[17] Nandan Nilekani, however, as we saw above, did not relate his company's singular investment in software development as unconstructive. He had specifically mentioned that they could have also worked on products, but it was something they decided not to pursue and that he is satisfied with Finacle and its success as the only product of Infosys. The fact that software development was not necessarily a choice but rather a function of global disparity remained unmentioned. This concern was further ratified in a meeting I attended, organized by the NASSCOM office in Bangalore between the then IT secretary S. Gowda and a handful of entrepreneurs dedicated to designing software products. At that time, there seemed to be a significant dearth of state support for the products industry.

Interestingly, the cost-benefit analysis of software development and the erosion of other fundamental disciplines acutely disturbed Ravindra. He referred to this uneasiness as "my worry kit":

> Pure academic interest is disappearing, for IT, for money. There is an artificial movement to one side. Even if this industry is to succeed then also we need good academics to train, but nothing is being done to restore the balance. There is a lateral pressure; if you are not in IT, then you are useless. But the creative element of society is getting disrupted and social respect for other professions are going away and simple materialism is taking place. In recruitment interviews I always encourage academics to continue in that profession but I'm compelled by business requirements. I am worried about it.[18]

For Dr. J. K. Suresh, IT bears an inherent epistemological contradiction for India, which he explained as follows:

> We have grown up with stories of pain and pleasure which is beyond the locus of the human world that lacks concepts of "self" and the

---

17. Personal conversation, 2004.
18. Interview conducted on October 4, 2004.

"other." It has more shades of gray. This is the nondigital view of the world, but IT disapproves of this. . . . In IT the load on our cognitive abilities is high since software is not tangible. But knowledge itself has become an important factor of production in various forms, like, licenses, patents, and wages. It has led to the death of the sciences but all these are thought as "progress." But is "progress" not an impressionistic term?

To follow up on his nuanced exposition, I wanted to know why Dr. Suresh chose to shift to IT at a somewhat later stage in his career. Dr. Suresh recollected that he moved because he was entrusted with KM at Infosys, which aligned with his interest in epistemic systems. "Besides," he explained, "despite its digital qualities, the appreciation of complexity remains in IT, like in the Indian knowledge tradition."

Dr. Ravindra, on the other hand, has a more pragmatic view about his move: He feels IT is gradually organizing every human activity. Therefore, rather than avoiding IT, it is imperative that one makes the best use of it. He compares the promise of IT with *Vishwarupdarshan*: *vishwa*, means the universe, *rup* is the composite of its various dimensions, and *darshan* is seeing. Therefore IT is at once a tool and a mode to perceive the entire universe in all its complexities. It is this possibility that encouraged him to move to IT but only as an educator and researcher in E&R. We encounter an affective contradiction with both Dr. Suresh's and Dr. Ravindra's analyses of their personal location in IT and the potential of IT in general. On one hand, they are drawn to the immense possibility of IT, but, on the other hand, they are also attentive that IT does overshadow other epistemological foundations. Such contradictions, I began to conclude as I spoke with many other software developers, also had a generational character. The majority of younger developers did recognize the shift; however, having no direct experience with the nation-building era, the shift to them was just "normal" as change is and should be. Some, on the other hand, held a romantic view of the past where life was slower paced that in IT. However, it was mostly the senior employees who displayed nostalgia for the previous phase of industrialization and who expressed a critical interpretation of IT as problematic. The link between India and science is thus now more nuanced than the previous experience during the era of nation building.

"Samudramanthan"

Dr. Ravindra refers to the success of the IT industry as *Samudramanthan*: "IT has made a huge contribution to India but as in *Samudramanthan*, toxins have also come up." *Samudramanthan* is a combination of "samudra," sea, and "manthan," to churn. *Samudramanthan* is an episode in the *Puranas*, the Hindu mythology text, where the immortality of gods against demons (*asuras*) was secured. In other words, *Samudramanthan* created the gods. The ocean was churned to dredge the ambrosia (*amrita*) that the gods strategically consumed to become immortal. But as this process begun, the first fluid that surfaced was the poison (*halahal*) that Lord Shiva drank alone. He, however, retained the poison in his throat, giving it a blue color (Shiva is also known as *Neelkantha*, blue-colored throat). Later, as the ambrosia appeared, the gods consumed it, denying the demons their share. Largely, *Samudramanthan* conjures an image of upheaval: the sea being churned, toxins being released, ambrosia sprouting, and finally the battle between gods and demons to attain immortality. The eventual victory of the gods calms the entire picture as demons are banished to hell.

Like all mythological elicitations, Dr. Ravindra's use of *Samudramanthan* reveals a structured engagement with reality. "One cannot overlook the fact that we have to work in a country that has a corrupt government, but there is corruption in the West too. But India is evolving; it has evolved more in the last fifty years." He lists "political corruption," "the disruption of social values," "growing individuality, wife somewhere, husband somewhere," "absence of deep relationships as in New York," "the mix of Bangalore is changing and there is nothing Bangalorean about Bangalore anymore," and a "shift in the tastes and attitudes of youngsters" as some of the toxins IT has dredged up.

The harm of the toxin is finally negated by the bliss of ambrosia that guaranteed the immortality for the gods and created heaven, the "happy" abode in Hindu mythology. Likewise, for Dr. Ravindra, "IT will bring about a positive change." However, he does not list the epistemological issue of IT he had previously mentioned as one of the toxins produced. Toxins, in this rhetoric, seem to take on a social character. It is not difficult to see the inconsistencies that dominate the realm of IT. Rather than looking at them as aberrations, it is productive to consider them as constitutive of a new industry that has taken on itself a task beyond its realm.

Here *Vishwarupdarshan* and *Samudramanthan* pave the way for the aspirational ideology of IT workers. It is relatively easy to see that these aspirations also have a hegemonic quality: it is embedded with the idea of a Hindu nation. Earlier I referred to this as the "Hindutva Lite," that is, a version of Hindu fundamentalism that avoids militancy yet gently upholds Hindu imageries to narrate an ideology. These mythical imageries align well with the Infosys-Janaagraha attempt to cleanse the nation of the toxin of corruption. The global perspective, on the other hand, enhances such cleansing by unlocking the world that lies beyond the nation-think. In a counterintuitive way, though, in reality, such religious imageries, as much as they are perceived to be unworldly, are in the end tied to consumption practices where money is at the center. That such consumption practices are dredging up a new concoction of toxins—new forms of social inequalities—is not only ignored but is also rendered invisible. Such invisibilization processes are not mere coincidence when one recalls my discussion in Chapter 2 about how urban planning was designed to promote a certain kind of lifestyle for a certain kind of people deemed economically and socially valuable.

Yet the question remains: How are we to understand this invisible formation? On the reverse, what kinds of subject positions are created for those who desire visibility and for those who cannot even dream to desire? Following Spivak (2000) one could also think of this as the "secessionist culture of Bangalore's Silicon City. . . . There are several thousand software managers in the city," she writes, "mobile among companies, who do not resemble the "average Indian," that impossible figure" (2000: 10–11). IT professionals at Infosys and volunteers at Janaagraha consistently emphasize the contemporary newness of their lived experience as the socioeconomic freedom to live virtually in the real place called Bangalore. Spivak reminds us, "The words 'virtual' and 'real' belong to an earlier semiotic. Every rupture is also a repetition" (2000: 11). This rupture too evokes that of the past because the Indian IT industry is still a service provider rather than the seat of capital in the global market; the global capital hierarchy is troublesome. Spivak calls this "capital-fractured in agency." Thus visibility in this context comes with a sense of subjection, which, like most of my informants, Ravindra does not perceive.

4

Travails of Time

A Deferred Present

The auto-rickshaw driver tried to locate the address for a while but in the end gave up. "Ille side-e nillispri" (Please park on the side here), I said to him in my basic Kannada. He pulled up to the side of the road, I paid him the fare, and stepped out. I looked at the nondescript buildings that stood in front of me and finally found a plain metal plate bearing the name Centre for Budget and Policy Studies (CBPS). I was here to meet the secretary of CBPS, Vinod Vyasalu. A narrow flight of stairs led to CBPS on the third floor. A middle-aged gentleman who I conjectured to be Vyasalu, who was sitting at the far end of the hall, said, "Come in. You are right in time otherwise we would have missed each other. I will have to leave to take care of some visa arrangement for tomorrow in an hour." As I walked toward him, he pulled a chair for me and said, "I am Vinod Vyasalu."

"Where are you going?" I wanted to know. "To Brazil," he responded with a curious smile on the corner of his lip. "Brazil!" I was intrigued because the road from Bangalore usually leads to the United States and Western Europe or even Singapore. This time, the smile broke into laughter. He concluded, "Because we all belong to the Southern club!" CBPS focuses on the issue of financial decentralization in the Indian administrative system. Vyasalu was going to Porto Alegre to attend a workshop that would compare the effects of decentralization in countries such as Cambodia, China, India, Indonesia, Vietnam, Bolivia, Argentina, Mexico, Peru, and Brazil. The workshop would also establish a common platform to exchange ideas about improving governance. "How is governance defined in the context?" I wanted to know. "You noticed my smile before. I am not supposed to question; it is something I have to go to having chosen to work in the development field. And do you know that Porto Alegre is our new reference?" He paused for a few seconds, then said, "Maybe tomorrow we'll have another one—it's never done."

Indeed, it is never done. The referents for India are historically malleable. Another interesting shift animates the practice of referencing since the early 1990s: it has now shifted specifically to cities. Thus one sees a comparison between Bangalore and Porto Alegre rather than between India and Brazil, because now cities are nodes in the global circuit. In a way, the "urban turn," as Prakash terms this shift, places cities above and beyond their respective countries.[1] Vyasalu was alluding to the changing nature of the references for Bangalore.

What a global city ought to become oscillates among various cities, which are considered to be superior in some respect at different points in time. In one instance, as I discuss in Chapter 1, if Bangalore is to become a world-class city, it will need to emulate Singapore. In another, a marathon was organized in the city because that is what global cities like New York do. At one time it also wanted to become Detroit after a minister of the government of Karnataka returned from the city.[2] Or it could be Porto Alegre, as Vyasalu tells me. As a developmental economist, Vinod Vyasalu recognizes the vacillation of this aspiration.

CBPS is a nonprofit organization, which primarily focuses on optimal budget allocation and budgetary processes of local governments. Let us look at the mission statement of CBPS:

- To undertake research by analyzing budgets of the state by specifically looking at different sectors of importance affecting the urban & rural governance on the lines of the 73rd & 74th constitutional amendments.
- To contribute insights to policy makers as well as elected representatives to empower them to cater to the needs of the people better.
- To constructively engage with the state and its functionaries about citizen rights and to seek to redress these through lawful means.

---

1. Available at http://preview.sarai.net/journal/02PDF/03morphologies/02urban_turn.pdf (accessed September 22, 2006).

2. The public works minister of the government of Karnataka at the time, H. D. Revanna, returned from Detroit in mid-2005 and promised to the public that he would try to make the Bangalore roads as navigable as in Detroit. On the contrary, in 2006 *Reason* published an online article titled "What Detroit Can Learn from Bangalore: A Booming City's Lessons for a Town in Decline," available at http://www.thefreelibrary.com/What+Detroit+can+learn+from+Bangalore%3A+a+booming+city's+lessons+for+a...-a0146345614 (accessed April 18, 2015).

- To take research results to people to ensure meaningful debate.
- To create awareness and sensitize its stakeholders by disseminating research findings through various media.

As we can see, there is a clear emphasis on democracy as an electoral process where elected representatives and the associated administrators are at the center of implementation of public policies. This is quite different from the way the BATF was assembled with no elected representative as a member and later with Janaagraha's reluctance to engage with "corrupt" politicians. At the time of my fieldwork, CBPS was working on municipal finances in Karnataka. The report, which was later published, upholds the relation between the electorate and their elected representatives on two grounds: "a. Those that they know to be the direct wishes of their electorate. b. Those that they *think* will be in the best interest of their electorate" (Rath and Rao 2005: 1). The authors further emphasized "issues of responsive governance and the ability of ERs (elected representatives) to play the double role of listening to the electorate, and communicating with them about policy issues are integral to the concept of representation" (ibid.). Yet they argue that one cannot think of local electoral representatives in isolation from the bigger administrative structure they are part of and in part subservient to, as well, such as the state government. The fact that a large water supply scheme was surreptitiously forced on the CMCs attests to the hierarchical nature of the state. This kind of nonconstitutional citizen involvement, as I argue, creates a convenient citizens' assent to further erode the power of local representatives. Rath and Rao agree, as well: "They (Government of Karnataka) have always found it cumbersome and perhaps even frustrating to go through the usual route of political legitimation of the projects by engaging with the ERs at the local levels and creating consensus" (2005: 4).

This kind of resistance to the emerging discourse of the market takes Raymond Williams's "structures of feeling" somewhere different. Williams writes, "If the social is always in the past, in the sense that it is always formed, we have indeed to find other terms for the undeniable experience of the present" (1977: 128). The question here is not just about the emergence of the Infosys-Janaagraha complex to promote neoliberal ideas of governance. Critics like Vyasalu and Pani agree that governance practices need to be reformed, but not outside the electoral process. Instead they argue that governance ought to be reformed within the frame of electoral politics to make the process democratically equitable.

Notes on Restiveness

Let me return to the practice of referencing. Though they both are references, Porto Alegre and Singapore are in many ways not commensurable. While

Singapore showcases what a "world city" ought to be despite its location out-side the West, Porto Alegre, as Vyasalu mentioned, is another city from the south with "world-class" aspirations. In this sense, the references are inter-woven. Borrowing from Saskia Sassen (2005), to think of cities connected in the global circuit only as financial centers misses the question of urban gov-ernance, which is crucial to those aspiring for such a status. We may want to think of this as a global helix, rather than a circuit, because it winds through an eclectic and sometimes, unexpected set of urban nodes.

My argument here is that the practice of referencing, therefore, matters more than the referents themselves. This is a matter of time, where New York, Singapore, and Porto Alegre can come together in a single temporal route of "progress" toward "modernity" and "development." However, why does a postcolony like India continually need external references? In this chapter I argue that referencing mobilizes a certain temporal imagination of India. It reflects a sense of urgency to continually dissociate from the existing form of the nation-state at any given point in time. The current dissociation is man-dated by the success of the IT industry as a unique opportunity to display Bangalore as not only the esteemed 'Silicon Valley of India' but also as a space that is neoliberal ready.

In introducing Brasília, James Holston observed, "The journey to Brasília across the Central Plateau of Brazil is one of separation . . . from underde-velopment to the incongruously modern" (1989: 3). This separation animates the aspirations of postcolonies like India, as well, where the city has emerged as the epicenter of this aspiration. The present of a postcolony is thus one of relentless separation, which is never complete or perhaps, can never be completed. The present quietly endures in the contest between breaking away from a troublesome "nonmodern," "unscientific" past and the eagerness for a "modern," "scientific" future that is different. Yet what constitutes the fu-ture as adequately different from the past is the very domain of contestation. Postcolonies are hence framed through a narrative of restiveness. A fidgety narrative that moves back and forth in hope of finding something distinct but seldom does. The postcolonial condition is that condition where the present is forever deferred. Edward Said argues, that it "is not only a disagreement about what happened in the past, and what the past was but uncertainty over whether the past is really past or whether it continues, albeit in different forms, perhaps" (1993: 3). The prefix "post" with "colony" for an independent nation, is an indication of the troubling uncertainty Said is referring here.

What remains obscure, particularly from an ethnographic vantage point, is what comprises the postcolonial. Postcolonies condense a specific geneal-ogy of time, particularly in its dispersions, folded within the nation. How do specific episodes, such as the success of IT, seen as a national success, sway the genealogy? For an industry that is in the business of perfecting the present and the future, how does one account for the past? How do the "captains of the new economy" like Murthy, Nilekani, and Ramanathan reconcile the past with their plans for the future? The genealogy, I claim, rests on a dual treatment of the past: negation and nostalgia. The dual mode recruits and denies the past selectively to assemble an imagination that aligns with a neoliberal future whether the state enables and emulates the market.

In India ways of knowing have deep-rooted colonial and socialist genealogies whose effects are still rampant, actively influencing the modalities of knowing—its practice, its process, and its "ways." The invocation of juridical structures, the intervention of the bureaucratic machinery, and the expansive effect of power have been central to the instrumental apparatus of knowing. However, knowledge production is manufactured by the context that is itself socially fractured and politically contested. This is further enhanced in an era of neoliberalization when "freedom" is entrenched within the rhetoric of the market as the freedom of capitalist consumption. Thus, in such statements of freedom, basic commodities like water must appear as a consumable good organized through a relation between the seller and the buyer.

We encounter an deep epistemic shift: the state is no longer the sole agent for the delineation, production, and distribution of ways of knowing. Civil society organizations, corporations, international donor agencies, and multinational policy consultants now play a prominent role in the epistemologies of postcolonies. India is not unique in its induction into the neoliberal paradigm. However, what does make the experience of neoliberalism somewhat distinctive for the country is the success of IT and India's related entry into the circuit of globalization on a somewhat differential plane. This induction has enthused the IT industry to mold India as a new nation aligned with the professed practices and expectations of the developed world: ethical versus corrupt, efficient versus lazy, reliable versus capricious, stable versus volatile. However, as I have been depicting, it is not sheer economics that impels this aspiration for change, but an intricate blend of temporality, imagination, affect, and practice.

1947 and 1991

While conducting fieldwork at Infosys and Janaagraha, I was struck by the dual temporal references of 1947 and 1991 that were frequently used to narrate India. Both years denoted a "separation." While 1947 was noted as the foundation of India as a sovereign nation-state free from colonial rule, 1991 launched India almost half a century later into the circuit of globalization as a potential player in geopolitics. As recent as a couple of decades ago, in the media and in public and private spheres, 1947 was the dominant temporal reference used to describe India. However, a competing temporal reference has emerged since 1991. Or as Mazzarella puts it, "globalization had, for once, a rather precise meaning" (2005: 36).

Postcolonial time, as Ashis Nandy suggests, is a "warp." He writes, "It is not perhaps a terrible liability that, in South Asia, though the future may not always look open, the past rarely looks closed. I believe that social and political creativity requires this capacity for play" (2002: 4). Here I am interested in the "play" between 1947 and 1991 and how it bends the genealogy of the nation-state. While 1947 and 1991 both depict the emergence of a new nation, the narration and connotation of the "new" has shifted in the intervening years. The years are employed to bracket off the intervening time to critique the "errors" of the socialist-redistributive and mixed economy model of the Indian state.

The nationalist narrative imposes a perceived unity of the nation in the form of what Benedict Anderson (1991) calls an "imagined community." Thus, the imagined community is in effect a collection of disconnected communities, which is seldom tantamount to the cohesive idea of a nation. As Homi Bhabha suggests, new narratives arising from the "recesses of the national culture from which alternative constituencies of peoples and oppositional analytic capacities may emerge" are important to "draw attention to those easily obscured" (1994: 3). Here I ask: Does a critique of the nation-state always necessarily arise from the "recesses of the national culture"? Or, conversely, how does one understand discontent that arises within a dominant group, such as the IT industry? What kind of politics does one encounter in such critiques vis-à-vis the state?

Neither the IT industry nor its critique of the nation-state are located in the recesses of the nation; however, it wants to institute an alternative constituency of people, the neoliberal citizens. In thinking that an alternative narrative can only emerge from the margins, from the "recesses," where "disgrace" supposedly resides, distracts us from the politics of disgrace that can be activated from privileged groups. Nothing in contemporary India is plausibly more privileged than the middle class, IT professionals being an exemplary group. As I discussed in the introduction, over the last two decades this privilege has activated a form of middle-class politics that seeks to "capture" the state to advance its own class priorities. Predictably the middle-class rhetoric, however, does not directly advocate class interests. Rather, it is about postcolonial "corruption" and the urgency to unshackle the burden of the colonial past.

Ann Stoler has argued that "from the vantage point of the postcolonial, the notion of a history of the present has strong resonance and appeal. Colonial architectures, memorials and archives and the scientific disciplines that flourished under the guidance of colonial institutions are dissected as technologies of rule . . . are . . . embedded in the habitus of the present" (2002: 163). The "embeddedness" of the habitus of the postcolonial is therefore not an opposition between the colonial and the postcolonial. Rather, the postcolony carries a persistent colonial trace. It is difficult to fathom where the colonial ends, and the postcolonial begins, if at all. It is even difficult to grasp if the postcolonial has indeed attained what it aspires to be: "modern." Furthermore, modernity in this context is a historically fractured idea replete with competing claims based on class, caste, religion, language, and ethnicity. When a nation comes into being mustering "another reason," as Prakash puts it, modernity is also critically linked to differing imagination of what the nation ought to be.

Dipesh Chakrabarty has usefully suggested, that "modernity is easy to inhabit but difficult to define" (2002: xix). The measure of modernity, Chakrabarty contends, lies in the measure of "nonmodern" and "backward." The antithetical assessment defers the modernity while it also defers the present; nothing seems to be modern enough yet the persuasion continues in some form. Chakrabarty also argues that social science cannot completely overcome the "historical and contingent differences between societies" (xxii). Further extending his observation, I will argue that seen through the lens of IT, the present of a postcolony is a new chase to somehow negate these differences.

Ethnographically, the force of the "modern" and the "backward" in everyday life fuels a sense of uncertainty that repetitively pulls the narrative of the nation-state in numerous directions. The uncertainty, as I see it, is not just about separating and cataloging the modern and the traditional. Rather it is about detangling a web one is caught up in from within. Thus when the IT entrepreneurs visualize emerging India as a space of modernity and prepared for global recognition, it is refracted through a specific reading of history that selectively aligns with what is "modern" and what is "traditional." For instance, Infosys always gave traditional figurines carved out of metal (especially bronze) as gifts to clients. Very often these were figurines of gods and goddesses drawn from the Hindu pantheon. Yet these were chosen because of their "good" representation of the "Indian tradition." While management highlighted the artistic value of such handicrafts as unique to India, what were obscured were their religious underpinnings. The uncertain relationship between IT's concept of modernity and such representations of tradition, I contend, is not a conceptual liability. Rather, it invites a different set of question about the tangle between the two notions.

The uncertainty is further reinforced in an era of neoliberal capitalism where the alignment between the nation-state, science, and human life is not unequivocal. Discussing the agenda for the twenty-first century, Michael Fischer argues, "These are not only matters of skilled manpower flows or of new technologies but how interpretation works, symbolic resonances are mobilized, passions are challenged, risks are leveraged and how things fit together . . . they are matters of new emergent forms of life, of new ethical plateaus, of new civic political contexts and deep plays that implodes . . . in arenas of new scientific infrastructure in which market, law, code and norms compete for hegemonic control over the rules of play" (2005: 55).

For a postcolony, emergences and emergencies are relentless; so what is possibly new since 1991? With IT, a new hegemonic middle-class control over the state is emergent from civil society evident in the ethico-political narrative that I discuss. It is not the transaction of the state and the civil society that is new. Neither is the transaction between the state and the business new. For instance, the Tata and the Birla groups of industries have been close associates of the Indian state, particularly during the nation-building phase. However, the latter relationship was one of mutual support, while the relationship between the state and the IT industry is a curious mix of entitlement and distrust. While the IT industry is reluctant to entirely acknowledge the role the state played to enable the industry to trade globally, the expectation that the state subsequently relinquish governance to the market is very high.

Here let us return to Nilekani's refrain: "We put India on the map." The relationship Nilekani draws between science/technology and India is not new, particularly for a nation that has aspired to be a space defined by science and technology since the nationalist struggle (Prakash 1999). Yet it seems that he is alluding to something different. At a glance, the difference is obvious: IT is the celebrated conduit to the "West." At another level, Nilekani's use of "we" signals a sense of ownership of the nation as a whole, a sense that has emerged from succeeding globally despite the constraints and stigma of a developing country. This discourse locates IT as an isolated occurrence rather than in continuum with the legacy of science in India as Prakash has suggested. Such derecognition is not an oversight. It is a studied amnesia that indicates a desire to break away from the past in the rush to carve a new nation.

What is the nature of the new nation, and what makes the new imperative? Finally, the story of IT in India becomes compelling not simply in mapping its changes and continuities with the past or its ethnography in the present. Also interesting is how history is carefully remembered and forgotten to make the IT success narrative compelling to free the nation from its onerous past. The IT narrative is most discernible in its new ways of referencing the "West," the nation, and individual subjectivities.[3] Even though the referents themselves have remained unchanged, there is nonetheless a perceived shift in the *mode* in which they are referenced.

A Divided Commute

Cities offer a valuable entry point for genealogical analysis. Spatial forms and their concomitant practices reveal the ebb and flow of time as various

---

3. I do not mean to suggest any kind of homogeneity of experiencing this "shift." Populations placed differentially in the socioeconomic structure undoubtedly experienced the "shift" differently; I allude to the fact that some kind of change was under way with neoliberalization that was experienced by most.

groups struggle every day to lay their claims to the city. As one looked out of the window, the bus ride to Infosys provided a useful reading in this context. As the bus headed out of Koramangala, an upscale neighborhood that marked the southern limit of the city, the surroundings gradually started changing. The quaint bungalows and the plush high-rise apartments yielded way to shanty roadside settlements and dilapidated concrete constructions on either side of Hosur Road. The presence of colossal billboards, rising high above derelict structures, made the urban inequity even more visible.

The billboards blissfully advertised real estate developments, gated residential complexes, new domestic and international car models, seasonal fashion outfits, and department store sale events, among other things. Interestingly a significant number of these advertisements, especially the ones for real estate, displayed images of "happy" white families smiling in front of suburban homes most likely in the United States. As the tea continues to be overboiled in the tiny snack shack, as the litter around piles higher, and as patience wears out at the bus stop, life below the billboard is oblivious to the life promised above. The billboards are for the emerging middle class who are in motion on Hosur Road either on the buses provided by the IT companies or privately owned vehicles.

The bus finally leaves the dusty and chaotic milieu of Hosur Road to enter the quiet and orderly world of the Electronics City. The bus ride to Infosys is not simply a regular commute to work. It is also an exclusive space for the production and dissemination of the eminent status of the IT industry in the wider society. The juxtaposition of the leather interior of the Infosys bus with everybody seated vis-à-vis the overcrowded public buses, where one would be fortunate to secure a seat during the rush hour, reproduces this hierarchy through bodily arrangements.

Manuel Castells developed the concept of the "dual city" as "the polarization and the segmentation of the labor force under the impact of the processes of restructuring (that) has specific spatial manifestations" (1989: 203). He extends Saskia Sassen's argument that the schism of the occupational structure is connected to the "processes of restructuring of capital-labor relationships" that on the other hand it also creates new jobs at every level. In Castells's Marxian analysis, the dual city has "dual processes" of capitalist production and class formation. It is "not simply the urban social structure resulting from the juxtaposition of the rich and the poor . . . but the result of simultaneous and articulated processes of growth and decline" (1989: 206).

Castells explained the "dual processes" in the "dual city" in the rise of IT. He identified the introduction of IT as a turning point that organizes labor and capital in an unprecedented manner he calls the "informational mode of development." IT enables "capital's bargaining position vis-à-vis labor by providing the management with a broad range of options" and enhances "labor flexibility" that "forces a reclassification of job categories to accord with new tasks" (1989: 189). The fact that information underlies "processes of production, distribution, consumption and management" is now a well-established notion. However, the informational *mode* is not bound to the production of the present and the future only. It is also necessary that it reproduce the past that it selectively retains. As I show later, the logic of software programming imbricates the past within present modes of processing information. Thus, the introduction of IT in capitalist production is not solely about streamlining the future but also the past. Further, in a city like Bangalore, the information economy has a role beyond the limits of software; it is embedded in the social as a new way of knowing.

The billboards are instances of the "informational mode." Like IT, their messages are cogent, precise, and adroitly conceived for a specific audience. Distant as the images on the billboard may seem in their skin color (predominantly Caucasian), lifestyle, and geographical location (the United States or Europe), together they signify the image of the "good life." Conceptually, "good life" or well-being in this case is a blend of excess wealth, global connections, ostentatious consumption, youth, and health. The images dissociate from the past image of well-being—government jobs, modest earnings, unostentatious lifestyles, and so on. The reference has shifted from England to America or from an English to an American way of life and an apparent prospect of class mobility borrowed from the latter.

Carol Upadhya and A. R. Vasavi argue, "Despite their apparently consumption-oriented lifestyles, software professionals tend to be rather conservative with regard to financial planning and goals, and they tend to use their money to pursue 'traditional' middle class goals in addition to spending on consumer goods. . . . But while the ability to invest in property and save money at a young age marks this generation off from the previous one, it is significant that the goals themselves have not changed (own your own house, plan for economic security, and invest in whatever is required for the family's security and upward mobility, such as children's education)" (2008: 106).[4]

Stories of class mobility, as I discussed earlier, are still rare in the IT industry; it is at best visible as a move from the lower-middle class to the middle class or upper-middle class. Research has also shown that most IT workers' fathers were college-educated and employed in what was then considered high-paying white-collar jobs (such as in government offices, banks, corporate firms); most of the mothers too had college degrees, and the IT workers themselves had received education at one of the premiere engineering institutes of the country and already possessed significant levels of "soft

---

4. Other scholars such as Leela Fernandes, William Mazzarella, and Purnima Mankekar have also explained the new middle-class phenomenon in contemporary India. Though they have not exclusively concentrated on the IT industry, their arguments are applicable. However, in many ways IT's success is the harbinger of such class transformations.

skills" (Krishna and Brihmadesam 2006). Very few had a lower-middle-class background—that is, where the father was employed in a blue-collar job or for that matter a clerical job.

The power of the mobility narrative lies in its American reference of well-being, which is seen as historically less knotty than the British reference. It offers a chance to overcome the colonial trace and activate a political space where Indians can now participate as equals. Does this indicate a normalization of the West? Or is it a normalization tied perhaps to a consumptive and sensory connection to the international brands now flooding the Indian market? However, whether the normalization is real or perceived is quite not the issue here. I think, we need to raise a different question: If the West still lingers prominently in the discourse of the nation-state, what is its new manifestation and how does it become available to us ethnographically?

Nostalgia's Emissary

Toward the end of our conversation Srinath Batni looked pensive. He blithely stared through the expansive glass windows in his office to the sprawling green manicured lawns and said:

> It did not always look like this. We started in a small building in Koramangala; it's still there. There were not enough tables and chairs. The place had all its problems. Once I remember we had to hold a meeting to decide how we can keep the mosquitoes out that would come in the evening. So we decided to put nets on the windows, which kind of took care of the problem for the time being. Now when you look at all these those days seem unreal. I am not sure if you would have come to Infosys to interview any of us in 1992.

Srinath Batni was then the director of delivery excellence and full-time director of Infosys (he stepped down from the board in July 2014). However, Batni is careful to point out that he was not a founding member of Infosys like Narayan Murthy and Nilekani; he was inducted into the board in May 2000. He told me, "I started as a regular employee, but in 2000 I was welcomed to join the board of directors, and to date my case is an exception."

Stories of the initial struggle of Infosys in the early 1990s were prominent among all the board members and the senior officials who had been with the company since its foundation. Hema Ravichander, then the human resources director of Infosys, recounted the "mosquito story" at a workshop I attended. The stories were nostalgic but also carried a sense of respite that the initial struggle was then over. Apart from the quotidian nature of the mosquito story, this dual affective state signals a conflict about how IT professionals conceptualize and negotiate the past. To comprehend this conflict, let us return to my conversation with Batni:

> SDG: I was talking to Mr. Dinesh the other day,[5] and, as most of the senior management here say, I think that this initial struggle of Infosys which I can understand is very precious to all of you and rightly so, needs to be documented in some form. It should not be lost.
>
> BATNI: Yes, yes, but you should also realize it is a historical document. See, for example, both you and me were born after independence, right? Now, for example, if my father says how it was during the British days, how demoralizing it was. Fine, we will hear him, two times, four times. But then what? Then we will say what are these things of the past that you are talking about; it is boring. You can document it, but what do you do with it? You can keep on telling but to the present generation it does not mean anything because the present generation has seen a different India.
>
> SDG: But the different India would not have been possible without the past events, say, the freedom struggle, which was not painless, as we all know.
>
> BATNI: It is true. But how many of us really appreciate our freedom fighters?
>
> SDG: Do you not fear that the struggle that you undertook as you have been telling me will go away?
>
> BATNI: It will go away. It is a natural thing. Two people may read and appreciate that but today Gandhi is like a god, to be worshipped. Most of us cannot even relate to him, he has gone beyond us, so it really does not make any sense. [*He noticed my brief silence and smiled in an effort to negotiate his own nostalgia.*] Let me tell you about an incident. In China, I was having dinner, no lunch, with a client.[6] He was an elderly man. His secretary was a younger-generation Chinese. At the end of the meal he had cleared his plate. But his secretary had some food left on the plate. Even I had some food left on my plate; I could not finish. He looked at our plates and made a comment. He said during the Cultural Revolution time there was not enough food for everyone to eat so everything was rationed including clothes. So what government used to decide, for adults this is the meter of clothes, for child this is the meter . . . so he has a habit of whatever comes on the plate to eat it . . . but the younger person did not see this. So if you keep

5. K. Dinesh was one of the founding members of Infosys.
6. Batni was also one of the directors on the board of Infosys Technologies (Shanghai).

telling stories of freedom fighters or Infosys it does not work. Even
I was not in Infosys before 1992. The founders can tell their stories
and I have heard stories of what they went through. I can never be
part of that life; it happens with the next generation.

SDG: But there may be people in Infosys itself who may be interested
in your story—why deprive them?

BATNI: Yes, you are right. I know some younger people here who have
asked me about this. They are very few. We can try to educate all of
them when only 10 percent care, 5 percent retain it, and 2 percent
only internalize that. We do keep talking about this in some form.
But my personal thing is when I see the reaction of my target audience from interest a few years back to boredom to now, "I don't
care," What can you do? That is the attitude. You have to stop.

SDG: Is there a solution to this that you may be thinking of?

BATNI: Not much. I think we have to let it go and maybe that's a better
decision for all of us here to look to the future and not what has
happened in the past.

The above conversation raises two issues: first, Batni's careful bracketing
off the past of Infosys as something irrelevant to the present; second, the
parallel he drew between Infosys's struggle and the freedom struggle of India.
The convergence of these two issues locates an intertwined past: the collective
past of the nation, of the citizens, and of Infosys. It is the past embedded in
troubling layers that makes denial necessary, yet it is not easy.

Nietzsche here reminds us, "Only that which never ceases to hurt stays
in the memory" (1989: 61). Rather than thinking of Batni's denial of the past
merely as a strategy to advance IT, I am interested in its affective quality
and how the present is offered to us. Following Svetlana Boym, "Nostalgia
tantalizes us with its fundamental ambivalence; it is about repetition of the
unrepeatable, materialization of the immaterial . . . nostalgia charts space
on time and time on space and hinders the distinction between subject and
object; it is Janus-faced, like double-edged sword. To unearth the fragments
of nostalgia one needs a dual archaeology of memory and place, and dual
history of illusions and of actual practices" (2001: xviii).

It is the double-edgedness of nostalgia that explains the narrator's effort
to render it as something useless ("You can document it, but what do you do
with it?") in the present. It is not difficult to see the utilitarian connection
Batni offers here given the basic market premise on which IT is based, where
deadlines and datelines are crucial for the business to continue and succeed.
However, it fails to explain the affect that underscores this nuanced treatment
of the past.

The uselessness of nostalgia and the memory of the initial struggle of
Infosys, as Boym suggests has a "slowing down" effect: "The study of nostalgia

inevitably slows us down. There is after all, something outmoded about the very idea of longing. We long to prolong our time . . . against all odds resisting external pressures and flickering computer screens. . . . Nostalgic time is that time-out-of-time of daydreaming and longing that jeopardizes one's timetables and work ethic, even when working on nostalgia" (Boym 2001: xix). The "slowing down" is antithetical to all that IT stands for. IT is a metaphor for "motion" (Feldman 2000). Despite Batni's desire to talk about the past, despite my proposal on conserving the past in some form, it is difficult for Batni to afford the stillness of nostalgia in a context that advocates motion.

Motion and Liberty

The kind of temporal "motion" practiced, symbolized, and prescribed by IT is purportedly different from any that the nation had previously witnessed. The future was luring and seemed attainable, but it was often difficult to settle on an understanding of the "past." I discovered there were other ways in which biographical pasts intertwined with the past of Infosys and the nation as in the case of K. Dinesh, a cofounder of Infosys and then the head of the Quality, Information Systems and the Communication Design Group.[7]

> SDG: Why don't you start by telling me how you came to found Infosys along with others?
>
> DINESH: I come from a middle-class background. My father was a high school headmaster with eight children. I grew up in a lot of smaller towns in Karnataka as my father was getting transferred from one place to another. Then I did my graduation from Bangalore University and then my postgraduate. That is in brief my background. I started working while in college. I really started my career as a sorter, sorting letters in a railway station with the postal department. My salary was eighty rupees a month. This was back in 1971–1972.
>
> SDG: How did you get the job with the postal department?
>
> DINESH: On merit because I had passed the SSLC exam first class. BSc also gave some value. But there was no interview.
>
> SDG: Did it have to be a local post office?
>
> DINESH: No. They select you for Karnataka and they can post you anywhere in Karnataka. But they train you in Mysore.
>
> SDG: Was there a particular reason you took up the job?
>
> DINESH: Yes, because my father had retired. None of the brothers were employed. I am the third son, so I took up the job to support the family. Then from there I was working as a cable inspector

---

7. Dinesh retired in 2011.

for the telephones; you know, here the post and telegraph are the same department. There was a good jump in salary and the type of job was different. Then I went to work in UCO Bank as a clerk for about two and a half years. The nationalized banks gave security and the pay was good, so it was quite glamorous. But soon I did not find it as interesting as I had thought it would be. And I still believed that I had a lot more to offer. At that time the computers were slowly coming to India, but I did not know anything about them.

SDG: Are you referring to the mainframes?

DINESH: No these were ICL machines from U.K. This was also the time, around 1976, when IBM had exited India. I joined the state engineering department, and there I worked for four years in the computer department. I had a bond with them, and I completed the assignment. Then I went to Bombay and joined Patni Computer Systems. That is where all the founders of Infosys were. We wanted to build a world-class company.

Unlike the other founding members, Dinesh's biography was an illustration of class mobility. Neither Narayan Murthy nor Nandan Nilekani spoke about their personal lives in my conversations with them. They both belonged to the middle class, where economic stability, quality education, esteemed jobs, and so on, were in a sense given. For Dinesh, on the other hand, it was a protracted journey of personal struggle that finally led to his current upper-middle-class status.

Dinesh's voice picked up pace as he started recollecting his transition to IT and the initial days of planning a world-class company:

SDG: How did you all come together? Do you remember the days when you started talking about this?

DINESH: A lot of things had happened. Mr. Murthy had left Patni, but he was asked to stay there. There was some talk but nothing concrete as yet. But the only thing that we had seen in each other is sincerity and hard work. We all had the vision that one can make wealth, despite being from the middle class, legally and ethically. In India we need ethical integrity and that's what we want to show to the nation.

SDG: So you are suggesting that something is wrong with the way things are in the present and something needs to be done?

DINESH: Absolutely. We cannot live like this. Things will have to change. We have initiated the process by building a model company; it is for everybody else to follow. We are still reeling under a colonial administration and mind-set.

SDG: What is the colonial mind-set? It seems like you are not referring
   to the British as such, but to Indians.
DINESH: Yes. To think like the British, that India is the government's
   property. But we are part of India and rightly so. We have to demand our rights as citizens.

One feature that was discernable in Dinesh's narration, which also coincides with the other cofounders of Infosys, is a sense of discontent. His
observation in the previous section—"And I still believed that I had a lot more
to offer"—is critical here. The employment with the state did offer economic
security for his family, but at a personal level they seemed stagnating. Dinesh
was also referring to a time of Nehruvian nation building of self-reliance
where the state was the primary and the credible employer. Yet, for some, like
Dinesh, it was a languishing workspace.

The induction of computers in the workplace made a difference in how
one thought about professional work. The swift digitized logic of the computer
inspired a new way of work and life. It accelerated thinking, working, imagining, and living beyond the limits imposed by the state on individual lives. In
a way it is the emergence of a new ontological paradigm. For David Harvey
this "condition of postmodernity" is underscored by what he calls a "time-
space compression": "processes that so revolutionize the objective qualities of
space and time that we are forced to alter, sometimes in quite radical ways,
how we represent the world to ourselves. I use the word 'compression' because
a strong case can be made that the history of capitalism has been has been
characterized by speed-up in the pace of life, while also overcoming spatial
barriers that the world seems to collapse upon us" (1991: 240). Broadening
Harvey's argument, I would contend that the time-space compression/
instantaneity complex spurs an impatience with life. The impatience defers
the present while also reconstituting it. One lives for the future, yet it is the
imagination of the future that propels the present.

In Dinesh's narration we find an instance of impatience particularly with
the insularity of the Indian state and its commitment to a socialist system.
Where did the imagination of nation building break down? What is the nature
of the discontent with the state apparatus? Does this indicate disillusionment
with nation building and a search for a different form of self-reliance? Is it a
search for an alternative model for the self and the nation?

Dinesh's use of "vision" to describe what brought the founders together
underscores the discontent and the dissociation realized in their ultimate
dream of founding a world-class company. Apparently, the discontent and the
dissociation were related to the profession; deeper, it was an ethical proclama-
tion. The IT industry was not only an opportunity that opened up new professional possibilities but also a political space for the assertion of supposedly
exclusive middle-class ethics: "integrity" and "diligence."

Deadlines and Datelines

It was August. My last week at Infosys was coming to an end. I recalled having coffee with Sukumar in 2002, discussing my research plans and future relocation to Bangalore in 2004. Sukumar was then heading the corporate planning division of the company. The following is a slice of that conversation:

SUKUMAR: How long do you think this research will take you?

SDG: Well, this time around I want to do some spadework for the project. I plan to come back next summer for some more preliminary work. And then finally in 2004 I would like to be here for about twelve to fifteen months to do the main body of the research.

SUKUMAR: This long? Well, I guess, every discipline is different. You know we often get management students from Wharton and Harvard and all those places they in fact complete their research in a month or so. They come over the summer break, stay here, and then they are done. But yours must be different. But I still cannot imagine that you will put in close to two years to do the research, when I think that we churn out software at the most in six months. I do admire your patience.

SDG: It's strange how we think of time differently. For me even two years do not seem enough, but for you it's too much time.

SUKUMAR: For you the time allotted for our projects is almost ridiculous, yet you are doing your work and we are doing ours . . . it is strange, but time is of immense value in our profession. [*He paused for a while.*] Sometimes I feel I am running a marathon of unending deadlines.

If motion and speed define and distinguish the IT industry, it is underscored by a *new* sense of time. It is new in two ways. The first one I will be addressing is technical in nature. Aneesh Aneesh explains the new time as "temporal integration":

The speed of electronic flows brings different time zones together and connects them in real time. Work is integrated across geographies, aided by the logic of programming schemes,[8] including information protocols that facilitate electronic flows through adaptive routing. These protocols periodically evaluate the fastest route between two points in the network, taking stock of the current traffic in the

---

8. Aneesh introduces the concept of "programming scheme" to understand "scripts of governance or productive control" in the current era of global capital. Programming schemes are shared by both state and nonstate forms of administration that control our ways of life, ranging from ATM machines and traffic models to socioeconomic behavior and beyond.

network, broken routes, and other problems. Guided by a routing algorithm, electronic packets hop from node to node, casting the net of real time over the globe. (2006: 84)

Aneesh highlights the path and the resolve that underline the passage of software code between India and the United States. The software program must reach the client within a stipulated time. Below I offer a conference call between Infosys and a client in the United States I witnessed to instantiate the nature of time in software development.

It was five o'clock in the afternoon, an unusual time for me to arrive at Infosys because I was usually there by eight in the morning. The buses had left for the city and the campus wore a tired look at the end of the day. But not for all. Nikhil Rajendra and his team would be spending the night to work with their client in Massachusetts, where the day had just begun. This team had been working on the client's project for the last two months, and the deadline for the project to enter the testing phase was next week.

I reached Nikhil's office fifteen minutes prior to the phone call as he had requested, "so that we are all settled." Nikhil was reading his email: "This is from the client. Every week before the call we exchange emails updating each other on where we are today with the project and they let us know if they detected any problems with the programs that we sent them. This way we do not waste time; the call time is usefully spent and priorities are set. I have also emailed him the meeting minutes from last week. I was waiting for the client's mail and now we can go to the conference room. Come."

"This is soundproof," Nikhil whispered in my ear as he opened the door, indicating the need for a lower voice modulation. Members of his team sat down around the table, close to the phone. The phone had extensive keys and switches. Nikhil looked at the clock on his cell phone. One of teammates quipped, "This is the most reliable time, because its satellite time. Nobody can play with it." "Are all the cell phones on vibrating mode?" Nikhil wanted to know. I checked mine and nodded. He turned toward me and said, "I will introduce you to the client, so that they know you are here. That's fine, right?" "Absolutely." The preparation for the call seemed complete, and Nikhil picked up the phone and dialed the number. Within a few seconds a male voice surfaced on the speaker:

NIKHIL: Hello, David, how are you?
DAVID: I am fine. How are the others?
[*The rest chimed in.*]: Fine, good, OK, thanks . . .
NIKHIL: David, we have a guest today, Simanti. She is in fact from the U.S. She is doing research for her Ph.D. here in Infosys. She is sitting in on this call to study how calls are conducted between clients and us. Hope you are OK with this.
DAVID: Hi, Simanti! I have no problem as long as this is for research.

SDG: David, thank you. I assure you it's only for my research.

DAVID: Where in the U.S. are you from?

SDG: I am originally from Calcutta but now based in New York and New Jersey.

DAVID: Great! Nikhil, I am reading the email as we speak. So the timeline seems to be working fine this far; we should be able to start testing as planned. But we are also thinking of adding a feature, an option, a log file. We are reviewing the design. We will be done end of next week. What do you say?

NIKHIL: What kind of time are you looking at for us to complete the add-on? Because this will need some more time, as you can see.

DAVID: Certainly, but not beyond two weeks, I'd say. It is just a small thing.

NIKHIL [*turning to Srinivas, one of his teammates*]: Is this doable in two weeks?

SRINIVAS [*smiles*]: Yes, if we can stay a little longer.

DAVID: Sounds good to me. Have you started testing on your end? What is the end date for you?

NIKHIL: About a week.

DAVID: And what about the dialogue synchronization? When will you have it?

NIKHIL: By next weekend.

DAVID: Are you sure?

The phone call ended after discussing a few more issues related to the project. The call highlighted two related issues: first, the temporal dimension of IT work that underwrites every unit of the software project; second, an instance of a direct exchange between Infosys and one of its clients bound together by a rigid timeline.

Software projects are broken down into smaller modules at the start. Subsequent to the weekly conference calls, the modules are broken down further and the work is delegated to an individual or smaller groups. Both David and Nikhil are acutely aware of the importance of time since the cost of a project is based on an estimate of requisite hours. During the entire duration of the call, Srinivas was documenting every word being uttered, which will be emailed as "minutes" which will be shared with the client.[9] Later, Srinivas showed me that he had specifically underlined the time frames that were being decided between David and Nikhil:

SRINIVAS: So later on there is no confusion. David will review the minutes and email back his feedback, and only then can this phone call be considered done.

---

9. The "minutes document," as it is referred to, lists the tasks to be completed in the coming week, priorities, and the timeline to which the two parties have agreed to adhere.

SDG: What if you miss a deadline for some reason? It can happen.
SRINIVAS: It's not an option. Honestly speaking, we cannot even think
    like that. And if we do, it better be for a good reason. Otherwise,
    a missed deadline is escalated; it will show up in QA as a "risk"
    project and finally may also reach Nandan Nilekani. Who wants
    all these troubles? It's better to put in a few extra hours and get
    the job done.

Recording meeting minutes is a standard practice of modern forms of
governance, both state and nonstate. It is in effect a normative measure, and
the IT industry is not different in this regard. However, the difference lies
in the kind of escalation that results from the temporal failure that Srinivas
mentioned above. As I have shown in Chapter 3, the QA department is the
conscience-keeper of the company. "Temporal integration" is therefore not
limited to the travel of software code. Extending Aneesh's argument, I'd
contend that fundamentally it is about infusing time with ethics. In the next
section I bring together a specific kind of time, that is the time of the nation,
alongside the new time conceived through IT.

Nation and Time

Benedict Anderson argued that nations can only begin through the bracketing off a "homogeneous time" that is "empty." He cites the use of calendar
and clocks as instances of the homogeneous time that assembles the idea of
a nation:

> The idea of a sociological organism moving calendrically through ho-
> mogeneous empty time is a precise analogue of the idea of the nation,
> which also in conceived as solid community moving steadily down
> (or up) history. An American will never meet or even know the names
> of more than a handful of his 240,000-odd fellow Americans. . . . But
> he has complete confidence in their steady, anonymous simultaneous
> activity. (1991: 26)

Bhabha notes that Anderson's use of homogeneous, empty time "misses
the profound ambivalence that Benjamin places deep within the utterance of
the narrative of modernity" (1994: 161). Anderson thus overlooks the "margins of modernity" where "we encounter the question of cultural difference
as the perplexity of living and writing the nation" (ibid.).[10] Reiterating my

---

10. Bhabha specifically mentions "writing" because, for Anderson, the novel and the newspaper, both forms of print capitalism, are crucial components through which the community of
the nation is imagined.

previous argument, the ambivalence of modernity does not necessarily lie in the margins of the nation. Further, the "homogeneity" and the "emptiness" of time raises issues of ownership of the nation. It belongs to all citizens, yet it belongs to none. It is thus malleable to different claims of ownership.

Ownership, in this sense, is an uneven and contested idea. In postcolonies like India, it is tied to the "obligation to forget" or "forgetting to remember" the past (Renan quoted in Bhabha 1994). Picking up my conversation with K. Dinesh here, I show how the question of temporality is negotiated:

> DINESH: It was not easy at all. We were thinking of building a world-class company in a country that is famous, no, in fact, infamous for corruption and an incompetent bureaucracy. Walk into any government office and all you see is laziness. Well, I have seen it myself and I found it just unbearable, and here I am spending my days usefully.
>
> SDG: How did you overcome the hurdles that you mention? Are they all related to the way the government works in this country? Or were there other kinds of problems?
>
> DINESH: Most of them are related to the attitude of the government.
>
> SDG: What about the 1991 reforms that surely helped the IT industry. Didn't it?
>
> DINESH: Yes, I know everybody thinks that the reforms can do everything, but they cannot. It is not about reforms but about telling the world that we are people they can trust. That is the most important thing. We had to undo what happened before and build trust, it was not easy and we did it by dint of hard work.
>
> SDG: Can you recount for me, the first few years of Infosys and the basics of doing work across countries?
>
> DINESH: We have used the links with multinationals that were exposed to other countries and other cultures. Like Reebok. GE [General Electric] in fact was a crucial break in the IT industry in India when they expressed faith in the domain expertise of India in this regard. We started by working in the U.S. mostly, but then slowly moved back our operation to India. It will seem strange to you now, but the actual software tapes would either be physically mailed or if somebody was going to the U.S., he or she would carry it. The feedback would be faxed to us, and we would again work on it. The point was to prove to the client that we can do the work; it is a process of building trust. Once the trust was in place, it did not matter where we worked from, like now. Twenty years ago, this was not easy as it sounds today. Even today clients have to be convinced about our capabilities. It definitely has decreased over the years, but it has not gone away.

The move from distrust to trust is an affective and a temporal shift. The foundation and success of Infosys is considered an index of this shift. The struggle is not merely economic. The desire to build a world-class company is not limited to technological excellence alone; it is also about ethical excellence. Extending Prakash's argument about the inseparability of science and India, IT adds a much-needed element to this imagination—ethics. The deliberate incorporation of ethics shifts the temporality of the nation-state from past marginality to future global recognition. IT and its concomitant ethical commitment accelerate India's prospects as, what has been termed, a "rising superpower." The claim to power is nonetheless shallow for a country where only a sliver of the population can participate in and afford this kind of rather grandiose imagination vis-à-vis the West.

A Normalized West

BATNI: So when do you head back to the U.S.?

SDG: Sometime towards the middle of August, with about two weeks to go before the fall semester begins.

BATNI: Too bad. I really wish you did not have to. You know now there is an option to live well in this country. India has changed quite a bit. You can have the same lifestyle if not better than in the U.S. There you have to do all the work yourself, and also drive your car; here you have as many maids as you need and also drivers.

SDG: Yes, it is true. You have lived in the U.S. and in retrospect it seems that you'd rather live here. Is it is because this is your "home" or is it for the convenience?

BATNI: I'd live here, but earlier we were not able to because the country had very little to offer in terms of what we wanted to do. So we were automatically drawn to the West in search of professional satisfaction and the reward for hard work. But now everything is available here, so why go elsewhere?

SDG: When you say, everything, what are you including?

BATNI: Whatever you want—Toyotas, malls, departmental stores, supermarkets, good restaurants, and whatever else you do in the U.S. Well, also good service from private operators, phone, Internet, everything.

SDG: So it is a lifestyle change.

BATNI: And more . . . listen to this. I was driving with a friend in the Bay Area; he has settled in the U.S. for a while now. We were both quiet. After a while he said, "Srinath, you guys are doing so much better than us," and believe me, he looked sad. I told him that he can now return to India, but nobody is very willing because they know that they have missed the bus. The other reason he tells me that I am better off than him is because Infosys is an icon of India today; it is a movement and I am part of that movement.

In a way, Batni normalizes the West by indexing the various everyday amenities and "Western experiences" offered by cities like Bangalore now. Normalization is also linked to homecoming, a reverse brain drain. IT made it possible to espouse a perceived Western life without leaving India. The experiences of the West—professional satisfaction, material comfort, and individual freedom—are now available in India.

When I was getting ready to leave Bangalore, several friends both at Infosys and Janaagraha expressed a sentiment similar to Srinath. They were sorry for me. First, that I had to leave "home" and second, that I would be unable to partake in the new possibilities that India now offered. Overall, in their opinion, I was *unnecessarily* depriving myself of the gratification of being home. "Don't mistake me," said a friend. "I'd love to go to the U.S. again, as I had two years ago, but not to settle there. It is only a résumé builder and, well, also increases my status with friends and family. You know where the prestige lies now? Not in living in the U.S., but going there on business and coming back to India." The travel back and forth, between India and the West, is perceived to be a path through which Indian IT professionals see themselves freed of the need to settle in the United States.[11]

Practical Patriotism

In *Modernity at Large*, Arjun Appadurai claims, "We need to think of ourselves beyond the nation." In this context, he raises the question, "Does patriotism have a future?" (1996: 158). However, I argue that the need to transcend the nation and the subject of patriotism are not necessarily contradictory. Rather, the apparent contradiction between the global and the national produces an unprecedented kind of patriotism. By unprecedented, I am indicating the form of patriotism, for instance, espoused by Janaagraha, which is a careful permutation of the past nationalist movement and the present neoliberal aspiration.

Appadurai conceives of the "imagination as a social practice," where "imagination has become an organized field of social practices, a form of work (in the sense of both labor and culturally organized practices) and a form of

---

11. Aneesh analyzes the divided existence of Indian software programmers between India and the United States in a similar context of the location of "home" in *Virtual Migration*. His analysis offers a nuanced and complex understanding of "home" where software professionals are emotionally and materially indecisive about where and how they ought to live. One of the best examples that I have encountered is how new residential real estate properties in the outskirts of Bangalore intended for U.S.-returned Indians replicate suburban American houses complete with family rooms, basements, and so on, otherwise unfamiliar architectural features in India.

negotiation between sites of agency (individuals) and globally defined fields of possibility" (1996: 31). The possibilities of the global, however, do not foreclose the nation. The global plays with the national imagination, reorienting and reinventing the nation-form in the process.

An instance of this play is evident in what Janaagraha called "Practical Patriotism."[12] Ramesh Ramanathan consistently described Janaagraha as a "platform for Practical Patriotism" and "Professional Volunteerism." In a meeting between Janaagraha and the Banshankari Welfare Association and Consumer Care Society, Ramanathan explicated the significance of "Practical Patriotism" as follows:

> For most of us, democracy is a romantic idea; but what we need is doing democracy. What is doing democracy? Ours is a great country with five-thousand-year-old history, but we do not think that we have a good government and every day when we read the papers we say, "Democracy is going to the dogs." Most people say I am honest. I am paying my taxes and governance is not my business. Such people should realize that democracy is an opportunity for us to engage, it is not a guarantee. People ask me how they can keep their job and family and still participate in governance. My answer is "Practical Patriotism." If the people could give just two hours a week, Janaagraha would show them how to use that time for making their ward the best. Many people say Indians are apathetic, but I haven't come across even one. I have come across people who have lost hope. That is not apathy. We have to bridge the gap between the potential of India and the reality of where we are today. Spirit of Community is the inspiration for Janaagraha.

Patriotism in the national memory of India is historically linked to the nationalist movement. Patriotism was emotive in nature, captured in celebrated aphorisms, like Bal Gangadhar Tilak's "Swaraj (Independence) Is My Birthright" or Subhas Chandra Bose's "Dilli Chalo (go)" to depose the colonial power. "Practical Patriotism," on the other hand, sounds like an oxymoron not in its rhetorical dissociation from affect, but in its blend of affect and pragmatism. Ramanathan dwells on affect when he alludes to the "great country" that is in need of revival and how it is important for "hope" to return among citizens.

Albeit its affective side, "Practical Patriotism" in Ramanathan's words, is also a quantitative idea—a "two hours a week" commitment that is equivalent to "doing democracy." This computation of labor in terms of man-hours

12. The first time I heard this term was when Ramesh Ramanathan was summing up the discussions of the Monday Morning Meeting very early on in my fieldwork in 2004.

is similar to software projects where part of the cost was based on billable hours. It was also standard practice at Janaagraha to announce the number of "man-hours" that have been devoted to a certain project. The definite temporal quantity belies the amorphous, unstructured, and idealistic nationalist sentiments and rebellious practices of the independence movement. Thus pragmatism in this viewpoint is not necessarily antithetical to emotion; it allocates and organizes emotion through structured time. Like the industrial time-thrift at Infosys, every minute of patriotic time too is meant to be productive.

The question of "Practical Patriotism" also brings us to Benedict Anderson's well-known observation: "What makes people love and die for nations, as well as hate and kill in their name?" (Anderson 1991: back cover). The stake in Practical Patriotism is not biological death, but one that brings death to the "corrupt" nation. Ramanathan once poignantly observed, "What we need is the death of the corrupt nation for a new one to be born." The reversal here is interesting—Practical Patriotism calls for the death of the existing nation rather than that of citizens. This death is linked to a corporate view of citizens as stakeholders of the nation and its singular emphasis on the economics of transparency and accountability. The question remains: How does the stakeholder view alter the imagination of the nation? If nationalism anchors on passion, what does Practical Patriotism attach itself to? Where does the perceived "deep, horizontal comradeship" of India that Infosys and Janaagraha envision now lie?

The following is an excerpt from an article by Ramanathan titled "What Is the Number?" in *Mint*:[13]

Beyond our [Ramanathan and his wife, Swathi] personal journey, however, what has been extraordinary in these past eight years is watching the changing relationship that Indians are having with money. From a time not too long ago when overt financial aspiration was frowned upon, it seems now that middle-class India is trying to make up for lost time. I suppose this is inevitable—we will become a developed country only when millions of individuals improve their financial position, across the social spectrum. This means the poor become less poor, the middle-class become rich. And a few become extraordinarily wealthy. . . . When it explodes onto a society's consciousness, money changes all the old rules of engagement. For instance, in my interactions with senior government officials, I wonder, "How can this person be expected to support liberalisation and private enterprise, change laws so that people can become millionaires,

13. *Mint* is a business newspaper published by Hindustan Times, a national news corporation, in agreement with the *Wall Street Journal* to publish their news in India.

and yet go home with Rs 30,000 a month? Won't his daughter also want Nike shoes and summer holidays?" It would be naïve to ignore the attendant temptation of corruption, and the moral consequences of the corrosion of character.

Ramanathan espouses the familiar trickle-down theory of economic prosperity. Though India has not witnessed a major trickle-down effect and the class disparities have grown even wider after liberalization, the rhetoric is still insistent. Also note Ramanathan's direct avowal of the middle class at the center of the new financial aspiration. Further, he cites a pair of Nikes as an index of prosperity and the ability to consume like members of the developed world. His reading of liberalization is closely tied to participation in the market and the middle class as the new consumers of brand goods that have lent the Indian marketplace global options (Fernandes 2006). For the state, the GDP provides the index of progress; for citizens, it provides their personal wealth. The new mooring of the "imagined community" lies in the active commitment to wealth creation and the related capacity to take part in the market as avid consumers. It is the aspiration to a certain lifestyle that binds the middle class together as the patriots of a new India.

Return and Sacrifice

Like Infosys, Janaagraha's rendition of the national past also involves simultaneous forgetting and remembering. The past is readily available, as I mentioned earlier, in Janaagraha's obvious association with Mahatma Gandhi's "Satyagraha." The accompanying adage "Be the change you want to see" is also borrowed from Gandhi. Unlike the origin stories of Infosys, that of Janaagraha is readily accessible in the public sphere. I collated the story as follows: Ramesh Ramanathan and his wife had returned to India, "sacrificing" their careers as a banker and an interior designer, respectively, in London (previously they had lived in New York). They had closely observed the ways in which Western civil societies and states function. The commitment of both the state and the citizens toward a good life vis-à-vis a fragile democratic system in India stimulated a passion in them to change India. The contrast seemed absolutely unacceptable to them after fifty years of independence. Unlike most nonresident Indians (NRI) who embrace distant nationalism, the Ramanathans decided to move back to Bangalore, their hometown, to "start what a friend called a 'no revenue model' phase to our lives"—a "citizens movement."

Bangalore has been the favorite destination of NRIs who wish to return to India from the United States or Europe mainly to continue with their IT jobs and a Western lifestyle. The city even has an association called Returned NRI (RNRI) to provide logistical help with housing, schooling, and so on, to those who are planning to come back. The return of the Ramanathans is in some ways considered unique if not unprecedented. Part of the reason lies in their commitment to change India as a nation-state rather than address individual social problems. Like Narayan Murthy and Nandan Nilekani, the Ramanathans wanted to address the fundamental problem of governance and citizenship as a way to redeem India. It enthused the media to compare Ramesh Ramanathan with Gandhi, who also relinquished his legal profession in South Africa to confront British colonial rule in India.

Ramesh Ramanathan has also inspired the storyline of *Swadesh* (home-land), a 2004 film starring the Bollywood star Shah Rukh Khan. Mohan Bhargava, the protagonist of the film, is a NASA engineer with an Ivy League degree who returns to India to find himself amid abysmal poverty in the village he visits. He commits to changing the plight of the villagers and successfully builds a hydroelectric unit for irrigation with his own funds. Ashutosh Gowariker, who directed the film, is renowned for making films that address nationalist sentiments. He had previously directed *Lagaan*, a film depicting a historical cricket match between British colonial officials and Indian villagers where the latter won. Based on the victory, the villagers were exempted from paying taxes, which was a relief given the severe drought that year.[14] The producer of *Swadesh*, Sunita Gowariker, was quoted in a national daily: "The main character in the film simply mobilizes people and puts them on the road to progress."[15] Ramanathan expressed a modest sense of satisfaction at one of the Monday Morning Meetings when volunteers commended him on his inspiration: "After all, people listen to Bollywood, so if they can make a difference using my story, I am all for it." Both *Lagaan* and *Swadesh* depict everyday struggles of ordinary people, earlier colonial subjects and later-day citizens—their imagination of a recovered life and their valor to confront the established structures of power. The unmistakable subtexts of the film, the media reports, and Ramanathan's own rhetoric are "sacrifice" and "social imagination," "change" and "public mobilization." The interest of Bollywood in capturing the return of impassioned NRIs in celluloid reveals its wider social impact not so much as a unique episode but as a possibility of changing India.

Between Gandhi and Nehru

Interestingly, the rhetoric of change espoused since the 1990s hinges on an opposition: it embraces Gandhi and denounces Jawaharlal Nehru, the first

---

14. *Lagaan* was also nominated for the Academy Award in 2003 in the Best Foreign Film category.

15. *Deccan Herald*, December 11, 2004, available at http://www.deccanherald.com/deccan herald/dec112004/c2.asp (accessed March 30, 2007).

prime minister of independent India. While Gandhi is now once again an integral component of the middle-class nationalist discourse, there is a troubling absence of Nehru. On one hand, the middle-class distrust of the state, a governance model introduced by Jawaharlal Nehru, is a statement against his vision of a self-reliant India. Often IT entrepreneurs and Janaagraha members argued that the Nehruvian model isolated India from the outside world, which negatively impacted the nation's progress and standing in geopolitics. On the other hand, Gandhi's "return," "sacrifice," commitment to "truth," and charisma to mobilize ordinary people for change were evoked regularly. The two leaders, it is known, endorsed different paths for India to follow as a nation-state vis-à-vis the question of the West and modernity. However, a close reading reveals that the distinction highlighted between the two nationalist leaders in the neoliberal narrative is not entirely tenable.

While Gandhi's imagination of India rested in the villages as self-sufficient units of economic and social production, Nehru believed in establishing science, technology, and infrastructure industries to set the country on its journey to modernization. I will draw on a correspondence between Gandhi and Nehru to elucidate their differences. On October 5, 1945, in a letter to Nehru, Gandhi wrote:

> I am convinced that if India is to attain true freedom and through India the world also, then sooner or later the fact must be recognized that people will have to live in villages, not in towns, in huts not in palaces.... We can realize truth and non-violence only in the simplicity of village life and this simplicity can best be found in the *Charkha* and all that the *Charkha* connotes. . . .[16] While I admire modern science, I find that it is the old looked at in the true light of modern science which should be reclothed and refashioned aright. You must not imagine that I am envisioning our village life as it is today. . . . They will not live in dirt and darkness as animals. Men and woman will be free and able to hold their own against anyone in the world. (Gandhi 2007: 150)

Nehru's response, dated October 9, 1945, followed:

> The whole question is how to achieve this society and what its contents should be. I do not understand why a village should necessarily embody truth and non-violence. A village, normally speaking, is backward intellectually and culturally and no progress can be made from a backward environment . . . we have to put down certain objectives

---

16. Charkha is a spinning wheel operated manually, primarily used to weave cotton.

like a sufficiency of food, clothing, housing, education, sanitation etc. which should be the minimum requirements for the country and for everyone. It is with these objectives in view that we must find out specially how to attain them speedily . . . there is no way out of it except to have them. If that is so, inevitably a measure of heavy industry exists. How far will that fit in with a purely village society? I do not think it is possible for India to be truly independent unless she is a technically advanced country. I am not thinking for the moment in terms of just armies but rather of scientific growth. (Gandhi 2007: 152–153)

These letters written in 1945 signify a critical year, that is, close to Independence in 1947. India eventually embraced the Nehruvian model of self-reliant state building premised on infrastructural work, dispensing with Gandhi's ideals of the village life. However, it is useful to recall Prakash's caveat on this difference: "While Gandhi's opposition to modern science and technology is well known, it is seldom recognized that Nehru was also a critique of Western industrialism. He shared in the running theme of Indian nationalism that class struggles and horrors of the industrial revolution, conjured up in Manchester, were alien to India, and that India should follow instead a path rooted in Indian soil" (1999: 203). The emphasis on the Indian soil explains Nehru's commitment to self-reliance as the preferred route for a postcolony to find its place in the world independent of the West.

The question that nationalists addressed on the eve of independence has resurfaced since the 1990s following liberalization: How should India progress as a nation-state? However, in its latest descent, the question now incorporates the issue of globalization as a force India *has* to acknowledge. Writing of historical descent, Michel Foucault argues, "Genealogy does not resemble the evolution of a species and does not map the destiny of a people. On the contrary, to follow the complex course of descent is to maintain passing events in their proper dispersion" (Rabinow 1984: 81). Dispersion is evident in the parallels drawn between Gandhi and Janaagraha where their ideological alignment is only partial. Ramanathan begins an article titled "Decentralization—The Second Wave" with a reference to Gandhi as follows:

Around the time of Independence, Gandhi said, "India lives in her villages," and asked that they become centres [sic] of self-government. Close to sixty years on, the argument is still valid—with one change: India no longer lives only in her villages. Already, we are close to 30% urban, and within the next 20 years, there will be more Indians living in cities and towns than in our villages. Responding to this challenge is all about strengthening local governments, but with some nuances. . . . Caught in the penumbra of the spotlight on their rural brethren, the urban dwellers are finding themselves in a governance vacuum, with all signs of the situation worsening.[17]

Ramanathan explicitly rejects Gandhi's ideal village life and actively promotes urbanism, which he argues has been vastly neglected. Janaagraha is unequivocal and unapologetic about its commitment to reforming urban governance and the freedom of the individual particularly to participate in the open economy. To promote Bangalore, the neoliberal center of India, is to promote an imagination of the nation associated with the expectations of the global circuit. Similar to Nehruvian nation building, the basis of this new alignment once again is a specific kind of supposed scientific rationale, that of the neoliberal market. The focus may have shifted from the Nehruvian mode of industrialization, but cities nonetheless continue as the formative location of the nation. The contemporary city additionally is also the active space of consumption, a discursive practice deemed necessary for India's integration into globalization. Overall, the science-rational-city complex continues as a Nehruvian descent in its creative dispersion to accommodate the market in an era of globalization.

Another striking parallel lies in social planning as a tool of governance. However, like everything else that is Nehruvian, the word "planning" does not appear in the Infosys-Janaagraha discourse, but reforming governance and citizenship was not devoid of a plan. Before we look into the neoliberal plan, let us look at what planning implied in the Nehruvian sense. In an essay appositely tilted 'Planning for Planning," Partha Chatterjee lays down the "significant aspects of the form of this exercise": "First, planning appeared as a form of determining *state* [sic] policy . . . the overall framework of a coordinated and consistent set of policies of a national state. . . . Second, [it] incorporated its most distinctive element: its constitution as a body of *experts*. . . . Third, the appeal to a 'committee of experts' was in itself an important instrument in resolving a political debate . . . about the central importance of industrialization for the development of a modern and prosperous nation" (1993b: 201). Chatterjee mentions the debate between Gandhi and Nehru in this context, but there is still a lack of consensus. "Rather," he contends, "the very institution of a process of planning became a means for the determination of priorities on behalf of the 'nation'" (1993b: 202).

The institutionalization of planning and prioritizing the needs of the nation vested power in the state and gave rise to what Bhikhu Parekh has called the "modernist programme" to battle "India's traditional way of life and thought." "The modernist programme for the regeneration of India consisted in creating and using a strong, interventionist, democratic, secular and

---

17. "Janaagraha, Advocacy," available at http://www.janaagraha.org/node/434 (accessed March 13, 2007).

centralized state to recreate society. Indians had become divided . . . and never thought of themselves as members of a single collectivity. They needed to acquire a common and uniform identity as citizens" (Parekh 1999: 67). What the IT industry and Janaagraha mainly critique is the Indian state's prolonged emphasis on centralized planning, which, in other words, they term "failure." They often argued that the failure does not lie in the tool of planning itself but in its exclusive location within and implementation by the state apparatus. Chatterjee argues that "the state was connected to the people-nation not simply through the procedural forms of representative government; it also acquired its representativeness by directing a program of economic development on behalf of the nation. The latter connected the sovereign powers of the state with the sovereignty of the people" (1993b: 203).

Chatterjee notes that the connection between legitimacy and the sovereignty of the state did not necessarily coincide and has been a continuing problem for the Indian state mainly in the rise of various secessionist movements from the margins of the state. The rise of the ethico-political critique of the state from the IT industry is another instance of this disconnect arising from frustration with the model of centralized governance. In the neoliberal discourse of planning, it critiques the state as the sole owner of the nation.[18] The concept of ownership is decisively transforming the rhetoric of the nation where both the state and citizens are equal stakeholders of India. The language of ownership is visibly embedded in the market rhetoric of who possess what and how.

Here, I think it is necessary to recall Raymond Williams's "structures of feeling" because this discussion brings together the "residual" and the "emergent" in a historical juxtaposition. By residual I mean the traces of colonialism, Nehruvianism, self-reliance, nation building, and so on; the emergent is the new sense of freedom ushered by neoliberalism since the late 1990s. Yet this juxtaposition is not merely about a collective in flux; we do see a certain structure that is already presenting itself, especially in the practices of the IT professionals and the feelings that underscore these practices. Williams writes, "It is a structured formation which, because it is at the very edge of semantic availability, has many of the characteristics of a pre-formation . . . it is thus

---

18. The focus of what mattered for sustaining and developing the Indian economy changed significantly in the 1990s along with liberalization. The following excerpt from the Planning Commission website is an indication of the change:

> The Eighth Plan could not take off in 1990 due to the fast changing political situation at the Centre [sic] and the years 1990–91 and 1991–92 were treated as Annual Plans. . . . For the first eight Plans the emphasis was on a growing public sector with massive investments in basic and heavy industries, but since the launch of the Ninth Plan in 1997, the emphasis on the public sector has become less pronounced and the current thinking on planning in the country, in general, is that it should increasingly be of an indicative nature. (Planning Commission, Government of India, "History," available at http://planningcommission.nic.in/aboutus/history/about.htm [accessed May 8, 2007])

a specific structure of particular linkages, particular emphases and suppressions, and, in what are often its most recognizable forms, particularly deep starting points and conclusions" (1977: 134). This is where we can map the semantic availability of a new ideology that brings together two otherwise unrelated domains—software and water.

The emergent semantic connection between the two is not yet complete, but the fact that even the state is struggling to hold on to the basic amenity discourse of water or is fairly reinventing itself to see water as *another* commodity is critical. At this time, the market seems the most viable option mainly because it suppresses state "corruption" and the political "noise" of an otherwise vibrant democracy, giving it a more legible framework. What we encounter at this ethnographic moment is a set of tension. It is with the diffusion of this tension at a later stage that a more generalized hegemonic description of the neoliberal social order will coalesce. Williams also reminds us about the "complex relation of differentiated structures of feeling to differentiated classes" (1977: 134). To tie this back to my foregoing discussion on the "middle class," I reiterate that despite the current rhetoric of class mobility and middle-class status as "achievable," the "structure of feeling" is also about how specific linkages are established that go beyond economics but are nonetheless interconnected, such as family background, prestige of educational institution attended, command of soft skills, familiarity with metropolitan life, and so forth. More often than not, the ability to speak English is highlighted as one of the main business advantages of the IT industry in India. What remains unaccounted for in these accounts is that knowledge of English is the baseline. In order to really move up the corporate ladder, one needs other skills, as well. The IT industry in this sense is breaking away from the class standards of the Nehruvian era, while also retaining the basic premise of middle-class sensibilities and at the same time displaying a new sense of self through a new set of semantics. Interestingly, Murthy and his wife, for instance, are still known for their "humble middle-class attitude and simple lifestyle."

5

"The Black Box"

The Black Box

The urban middle class in contemporary India is in revival mode. After the nationalist movement, the middle class is once again, I argue, following Chatterjee, "engaged in a pedagogical mission in relation to the rest of the society" (Chatterjee 2002: 172). If anything featured with unfailing regularity in the citizen participation meetings at Janaagraha, it is the pedagogical approach toward the urban poor: How do we educate them about the Water Project? Stated differently, how does one inform the poor about the value of the market? Like the general middle-class rhetoric, Janaagraha interpreted the lack of formal education among the urban poor as their obstacle to grasp notions of citizenship, governance, and, above all, the forces of globalization that have created a new urgency for India to accelerate its standing in the world. The rhetoric created a condition for excluding the urban poor most visible in the citizens' meetings organized around the city, which were overwhelmingly attended by the middle class. Even after several rounds of persuasion, the residents I knew from my work in Manjunathanagara seldom attended these meetings.

Ramanathan called the urban poor issue the "black box." The "black box," he would often explain, "stores important information crucial in the aftermath of an air crash to trace the last minutes leading to the disaster." "Black box" was sometimes used in reference to closed-source software

programming as a cognitive framework in Infosys. In the case of closed software (as opposed to open source, and this is the most intellectually and ethically debated phenomenon of the industry[1]), the source code is not available
to the user, but it can be independently embedded within other compatible
programs. The metaphorical use of the "black box" in the case of Ramanathan
was somewhat different from that of software. The black box in an airplane
records incoming and outgoing messages, and the relation that he builds
between disaster and the urban poor is significant. In other words, like the
black box, the urban poor we know are present but their presence need only
be acknowledged from a distance. This disengaged rhetoric of assigning the
urban poor to the realm of the black box is tantamount to relegating them
to a domain that *can* remain unknown. The lives of the poor become worthy
only in the context of a disaster, though Ramanathan never quite verbalized
what disaster meant in this case. Above all, this disengagement normalizes
the politics of class difference, where the middle-class rhetoric brackets off
poverty as an unchanging, mute, and inert reality.

I gathered that when Janaagraha was founded, Ramanathan took a keen
interest in the urban poor when he would make regular visits and hold meetings in the slums. However, by the time I was doing my fieldwork, Janaagraha
had mostly abandoned these meetings. I wanted to know why. Ramanathan
mentioned that during one such public meeting in the central quad of the
Nellorepuram slum, he came across a young man who had killed himself in
the early hours of the morning by hanging. "His body was still hanging from
the ceiling fan," he said, "waiting for the police to arrive. Except me, nobody
seemed to be that concerned. They said they were sad, but they wanted to start
the meeting. But I just could not. How could I? There is this body next door.
Nobody knows when the police will arrive; I could not speak. I felt miser-
able. Is this how little they value life?" By the time he finished, his eyes were
glistening with tears. "I tried," he said. Other volunteers who were there with
him that day corroborated his account of the troubling scene. He then pulled
back emotionally, insisting that though this was not a watershed event, "it did
bother my middle-class sensibilities." Consequently Janaagraha hired a staff
volunteer, interestingly from a lower-middle-class background, to visit the
slums and update them weekly on the urban poor situation.

Later, when I was in Nellorepuram to attend a water meeting, I spoke to
the deceased man's family and neighbors. The reaction I received was similar
to that Ramanathan mentioned: apathy. Ratnamma (name changed), one of
my primary collaborators in Manjunathanagara, once reflected: "We cannot
afford to cry too much over deaths or be too happy when somebody is born.

1. Open source is upheld for furthering innovation and adaptability in software programming.

They just happen and we are OK with both. When we are scrounging for food and water, there is no time left for these things. These are luxuries." The disaffection—the inability to agonize over death—appeared emotionally base to Ramanathan, whose upper-middle-class privilege normalizes and normativizes manifest mourning. This is not unique to India; Nancy Scheper-Hughes's *Death without Weeping* and João Biehl's *Vita* offer similar ethnographic analysis. Disaffection in this context is a politics of self-disciplining one's emotions vis-à-vis immediate corporal needs, such as hunger and survival. Nonetheless, many slum residents also affirmed that disaffect, as a strategy of survival, is difficult to commit and adhere to at all times. They emphasized that "because the rich feel we are not like them, that is, we are not even capable of emotions, they want to help us, and that help I can tell you is very important for us because one cannot rely only on the government these days. But they also do not think of us as human beings."

In all the meetings Nilekani attended or presided, he insisted that public governance should closely emulate the principles of corporate governance of accountability and transparency because of their inherent ethical thrust. Ramesh Ramanathan, endorsing these values regularly, stressed that there was nothing "corporate" about them; they are universal human values that all should follow. These values were also deemed "universal" and would be valid across classes, except that it was the "responsibility of the middle class now to evoke them once again for the entire society." It is at this ideological juncture between Infosys and Janaagraha, that the implementation of citizen participation in the Water Project became critical to "doing democracy."

In this chapter I address the middle-class politics that underscored this notion of participation and the construction of the urban poor 'problem' or as Hansen calls it the "hopeless urban melancholia" (Hansen 2001). I emphasize how this "problem" further marginalized the urban poor and their habitat—the slum—and the kind of resistance it evoked. Like in other developing countries, slums in India have been the object of liberal curiosity, detached humanitarianism, touristic voyeurism, and inane entertainment as well for the Western audience, popularized by Oscar-winning films such as *Slumdog Millionaire*. Mumbai, where *Slumdog Millionaire* was set, is often seen as the paradigmatic city to explore poverty, the slum, and the lives of slum residents (Gandy 2008; McFarlene 2008; Anand 2011). Overall, slums and slum residents are seen as problems, aberrations, and delinquents, among other things (Hansen 2005; von Schnitzler 2010). They are not considered citizens as members in the "homogeneous social" order; instead they are in the motley of "heterogeneous" populations (Chatterjee 2006). Yet they are a population in the margins of society, necessary for the state to exercise unbridled authority and use them as unsuspecting target populations for developmental projects. They can also act as vote banks in the electoral process.

Colin McFarlene has summarized the role of slums in her study of the Slum Sanitation Program (SSP) in Mumbai as follows:

> While the SSP is a regulatory intervention, it is important to note that because the "slum" inhabits a grey area between legality and illegality, and because of its contentious focus in debates about public space, state (and private) interventions are often just as likely to take the form of violence . . . [thus] we see both welfare-oriented interventions such as the SSP on the one hand, and violent acts of state demolition on the other . . . together they constitute the changing and often unpredictable field of relations between state and informal settlements. (2008: 99)

The citizenship claim that slum residents can lay on the state is therefore minimal; instead it is their reliance on spasmodic and unpredictable state benevolence in forms of projects like the SSP or the Water Project that offers a chance of full membership in the polity. On the other hand, as a negotiation strategy, the currency of referring to the state as *mai-baap* (mother-father) is particularly strong in the vocabulary of the poor. *Mai-baap* is not rights based but relies on the affective response of the state as a parental figure. Participation in public projects then could be seen as path to meaningful citizenship.[2] However, any meaningful participation of the poor is often thwarted by the educated middle class who perceive them as a monolith lacking civic consciousness or as the condemned subaltern (Chatterjee 2006) who are uninitiated into the juridical life of the nation-state (Das and Poole 2004).

Water and Manjunathanagara

It had rained all night, yielding a surprisingly clear and crisp morning. I was heading toward Greater Bangalore. The auto-rickshaw driver carefully negotiated the potholes filled with muddy water to avoid any damage to his vehicle. The concrete road soon ended and what lay before us was a meandering stretch of soggy mud. As I had expected the auto-rickshaw driver let me off, declining to go any further. I paid him the fare and, cautiously balancing on a few solitary bricks laid over the mud at random intervals, entered Manjunathanagara. Manjunathanagara was a slum with a population of 2,250 people living in about 450 tenements and would be included under the Water Project. It is a collection of shanty dwellings, referred to as "rooms" rather

---

2. Lawrence Cohen (1999) has made a similar argument about kidney donations by the poor. In order to donate a kidney, one must undergo a sterilization procedure, which is meant to curb population growth. The bodies of the poor are thus "bio-available" as gifts to the state in expectation of formal recognition.

than "houses." Most of them had tin or tile roofs; some were covered with plastic to keep out the rain. Few had been reconstructed with concrete, while most were built with a mix of clay and concrete.

I had started visiting Manjunathanagara after I met some of the residents at a meeting at the Water Board who were speaking in Hindi alongside Kannada. After the meeting, I introduced myself to the residents, who had gathered around a tea stall on the sidewalk bordering the Water Board, and asked them if I could visit the next day. I admitted to them that my Kannada was not very strong, but they assured me, as I had anticipated by their use of Hindi previously, that their slum was primarily composed of migrant laborers from Bihar, one of the northern Indian states. I had gone to several other slums that were quite ethnically and linguistically homogeneous. They were predominantly Tamil or Kannadigas because to secure a tenement in a slum, one had to tap into kinship networks based on linguistic ethnicity. In Manjunathanagara, the linguistic divide was not between Tamils and Kannadigas, which I mentioned earlier, carry a long historical strife in Bangalore (Nair 2005), but instead between the north and south of the country.

A narrow alley wound through the slum, lined with storm water drains on either side. In the absence of a regular cleaning schedule, the drains were clogged with solid waste all year-round. The condition worsened during the monsoon season when there was significant waterlogging and the waste rose from the drains to flood the walkways. Most of the houses were thus elevated approximately two to three feet to keep out the logged water and waste. There was no water supply within the slum, except for a public stand post (faucets) located at the entrance where water was supplied once a week by the local Mahadevapura CMC. The households also lacked any kind of lavatory facilities; women used public toilets built by an NGO and the men usually used an adjacent field.

Manjunathanagara was an illegal settlement and was not listed by the Karnataka Slum Clearance and Improvement Board (KSCIB). The Hindustan Aeronautics Limited (HAL), a para-statal organization that researches and designs defense aircraft for the government of India, legally owns the land. A city NGO had been actively involved in securing a legal status for Manjunathanagara with the KSCIB that would endow residents the right to basic amenities, particularly water.

One particular day the bustle of voices, mostly of women and children, seemed busier than usual. As I took a turn along the narrow pathway, parting my way through the clothesline, the reason soon became obvious. There was water flowing from the faucet! The women and children had gathered around the fountain, placing their pots and large bowls in a queue to collect water. These pots, which were traditionally made of clay, are now made of plastic. Some of these pots were terracotta color, reminiscent of their earthen past. Besides longer durability, the plastic reduces their weight and makes them easier to carry when filled with water. The women delicately balanced the pots on their hips with one hand around the necks.

"Was this scheduled?" I wanted to know from a woman standing next to me patiently waiting her turn. "No, but it is never planned; it depends on the mood of the valve man." There was a tap on my shoulder and it was Ratnamma with her usual welcoming smile. "When did you come?" "Just now. Where are your pots?" I wanted to know. "There." She pointed to the colorful array of pots amid the women, keeping close vigil on those being filled and moving up the next empty one below the fountain.

There was a noticeable but not unusual absence of men around the fountain, except for the valve man. The valve man is employed by the local municipal administration and is responsible for releasing the valve to supply water to the public fountains. Theoretically every slum should receive water once a week on a rotational basis. However, in everyday practice, the role of the valve man was more than administrative. He occupied a position of power vis-à-vis the slum dwellers where he could either withhold or release water indefinitely. The valve man's decision was mostly centered on financial and material incentives he expected and received from the slum residents. It also related to interpersonal relations such as kinship, friendship, or romance. Overall, the basis of the relationship on which water was supplied was tenuous, spasmodic and fragile. The faintest tension between the valve man and the residents upset the water delivery schedule.

Manjunathanagara was not scheduled to receive water on this particular day of the week. Nor did they receive water in the past week. Two weeks ago the valve man, Srinath (name changed), had gotten into an argument with some of the residents over a sum of money he had been demanding in order to release the weekly allocation of water. The problem seemed resolved after Srinath agreed to accept a lower sum than he had originally demanded. "He is also interested in one of our girls here," Ratnamma whispered in my ear. Srinath also agreed to supply water at a "decent" time in the morning, around seven, rather than in the early hours, which woke the residents from sleep and sent them into a zombielike scuttle to collect water.

The unanticipated agreement gave way to much-needed relief after almost a week of water hardship. During the past week Ratnamma and other women had walked nearly 5 kilometers (approximately 2 miles) round trip every other day to collect water from a neighboring public fountain. Women, sometimes those pregnant or sick, and adolescent girls were primarily responsible for securing water for the entire family. Even water usage was gendered: women tended to use water sparingly to make it available to the men in the family, who often used it liberally.

I started walking back with Ratnamma to her room after she filled her pot. She walked in her practiced steps, careful not to spill any water. As we reached her place, she pushed the door open, placed the pots in a corner, and settled down on a bamboo mat, gesturing for me to sit down beside her. "Do you want some lunch?" she asked. "No, I'm fine." "Then I will get you a drink." Before I could refuse, she called out to her son from the door and gave him a five-rupee bill to get me a bottle of Coca-Cola from the neighborhood convenience store. The host usually offers a drink, normally water, to the guest during the summer months, which was also ritually followed at Infosys. Despite my strong ideological differences with the company, Coca-Cola became my usual drink while working in Manjunathanagara. Ratnamma and the other women I knew there vehemently prevented me from drinking the water they themselves consumed. Nor did they allow me to pay for the Coke, which culturally would be tantamount to humiliating a guest. "We filter the water we drink seven times through a piece of cloth, yet its color is brown, like the cover of your notebook," one women explained to me in the beginning, "You just cannot drink that water; you will fall sick." Ratnamma settled down on the mat with a bowl of rice and *sambhar*, lentil soup, while I poured some Coke in a glass to share with her son.

The "Urban Poor" Conundrum

As I mentioned above the first time I met the residents of Manjunathanagara was at the Water Board. Along with other 'stakeholders' they were attending a meeting to discuss the "urban poor" module of the Water Project. However, the Water Board Act did not mention "urban poor" as a separate category, which created a sense of urgency among the sponsors of the Water Project, IFC and WSP of the World Bank. A WSP staff member mentioned to me, "We are in these projects only for the poor, and if BWSSB does not even consider them, how can we justify to Washington, D.C., to fund this project. It will really make us look bad; we need the poor." The fact that the Water Board has historically supplied free water through public fountains to most slums in the city was dismissed as an "informal arrangement." The "urban poor," an IFC personnel demanded, had to be "correctly," that is, juridically, instituted as a "special category."

The Water Board engineers and administrators received the censure of this "glaring omission," as surmised by an IFC staff, with a fair measure of resistance. As Srikanth Iyengar, a senior engineer, later complained to me, "Everybody, like the WSP, IFC, and WaterAid, are looking for placeholders, but BWSSB (Water Board) has been supplying water to the poor even before they were born. So if they need us to add urban poor as a category, we'll just add it in the Act; that's it." However, the Water Board was also aware that the inclusion of the urban poor as a category is not a simple juridical matter. It was not clear if the criteria to identify the urban poor would be solely economic, spatial, or both. It is very plausible, as was often argued in the Water Board meetings, that "not every urban poor lives in the slum and not very slum dweller is an urban poor." Also, Bangalore had been experiencing a surge of migrant laborers from the northern Indian states to work on the ever-expanding construction sites, and this rural-urban migration across states had further complicated the category of urban poor. Most of the migrant workers lived on the construction site till the project was over and then moved to occupy sidewalks in makeshift shelters, availing water at the free public water fountains. The Water Board termed them the "floating population." The "floating population" proved another challenge since they were considered squatters who would eventually have no access to water as public fountains were planned to be discontinued as part of the Water Project.

The urban poor issue was first addressed systematically in this meeting, titled *Water Project Stakeholders' Meeting to Discuss the "Citizen Report Card on the Public Services to the Poor in Peri-Urban Areas of Bangalore."* The chairman of the Water Board, representatives from WSP of the World Bank, IFC, WaterAid, Janaagraha, Public Affairs Center (PAC), and Association for Promoting Social Action (APSA) attended this meeting. The residents of Manjunathanagara were representing the "urban poor."

The Citizen Report Card, a tool developed by the Public Affairs Center, is described as

> an aggregate of public ratings on different aspects of service quality, built on scientific random sample surveys of users of different public services (utilities) in a city. The specific aspects addressed in the survey include availability of service, usage, satisfaction, service standards, major problems with service, effectiveness of grievance redress systems, corruption encountered and other hidden costs experienced by citizens on account of poor service.[3]

Sita Sekhar from PAC explained that APSA used the Citizen Report Card to assess the state of the water supply to the urban poor in the Greater Bangalore area. APSA had conducted research in four slums located in two CMCs, one of them being Manjunathanagara. The Manjunathanagara residents sat quietly at the end of the long mahogany table, looking around the boardroom with an inquisitive eye, sometimes sharing a word or two with their friends. Later when I asked them if they were familiar with the notion of the urban

---

3. This statement was available during the time of my research. Though the notion has not undergone any substantive change, the Citizen Report Card is now described as follows: "The Citizen Report Card (CRC) is a simple but powerful tool to provide public agencies with systematic feedback from users of public services. By collecting feedback on the quality and adequacy of public services from actual users, CRC provides a rigorous basis and a proactive agenda for communities, civil society organization or local governments to engage in a dialogue with service providers to improve the delivery of public services." Available at http://www.citizenreportcard .com (accessed April 20, 2015).

poor, they were not since it was in English and the word did not translate well into either Hindi or Kannada. The word "garib" in Hindi, meaning poor, does not capture the nuances of being poor in the urban context and its implication for development projects. As one resident apathetically said, "We do not need to, because they know everything." The power inequality in the room was deep: while the residents' presence ratified the "urban poor," they themselves were unable to partake meaningfully in the discussion in a context that relied not only on the English language but also on a vocabulary that was esoteric.

Neeta, representing WaterAid, started the discussion about her NGO's commitment to ensure water supply based on the three guiding principles of "equity," "access," and "citizen participation." However, she stated:

> Based on our experience of working in the water sector, citizen participation has been limited only to the middle-class section of the society and at this point we are working towards expanding the idea to include the poorer sections, as well. In our view with increasing liberalization the poor are increasingly getting marginalized and WaterAid intends to help them voice their specific concerns.

APSA followed three methods prescribed by PAC: focused discussion groups, social mapping, and a survey of 297 households. In their final report titled "Are They Being Served?" APSA presented to all the stakeholders the difficulties the urban poor face every day with intermittent water supply and want of underground drainage. The major findings were listed as follows: drinking water is available primarily through public taps, most slum residents walk to a nearby public stand post to collect water for the family, more than 60 percent of residents were satisfied with the water supply except that the timings were sometimes inconvenient, and, finally, the Willingness to Pay (WTP) survey indicated that most residents were keen to pay for individual connections. The chairman conceded that these are not abstract concerns but are deeply rooted in the everyday lives of the slum residents. He added, "But as you all know this is the first time BWSSB (the Water Board) is responsible for supplying water to the Greater Bangalore area, where we have not supplied before." He did not mention the bulk water supply to the IT Parks in the area, perhaps because it is a different arrangement.

The discussion then focused on the technical options of individual or shared connections that Water Board was planning for the slums. Sita Sekhar mentioned to the chairman that slum residents "are more willing to pay for individual connections rather than shared connections." PAC had used the WTP survey to gather this information. WTP is a survey method that solicits the citizen's answer about the present cost incurred to avail water and how much he or she is ideally willing to pay for improved and sustained levels of service in the future. The chairman responded that "BWSSB (Water Board) is flexible either in terms of individual or shared connections and that is according to what citizens decide is best for them."

The "readiness" of the urban poor to pay for individual connections generated some sense of relief at the Water Board following these presentations. Since the urban poor have historically been provided free water through public stand posts, there was some unease at the Water Board to now collect money for water. Some of the questions that preoccupied the Water Board administrators for a while were: How would they charge money for a basic public good such as water? How does one explain the benefits of the market to the urban poor given their lack of education? Will it provoke interference from political parties for whom the slums are a major vote bank?

The CRC report resolved these issues to some extent: The demand for individual connection indicated the WTP for water originates from the urban poor themselves. It lessened the dilemma that the Water Board was facing about charging the poor for water, because, as one administrator told me, "Well, now we can say they asked for it." The onus of participating in the water market thus shifted to the individual, who seemed "eager" to be integrated with it. However, the methodologies of evaluative tools such as the WTP and CRC that help in the creation of the market were exempt from scrutiny given their supposed "scientific" basis. Even the survey was conducted as a matter of protocol, which would make the apparent consensus doubtful. In the meantime, the Manjunathanagara residents sat quietly through the meeting. The tea in the expensive china turned cold in front of them till the server removed those. Later one resident mentioned, "I had not dared take a sip. I was scared I may break it." Unable to follow a word in the room, finally Ratnamma nudged the APSA volunteer sitting next to her. The volunteer listened carefully to Ratnamma's whispers, then raised her hand. "She has something to say."

The chairman, who was from the eastern state of Odisha and did not follow Kannada, said, "Let her say in Kannada and you can translate it for us." Ratnamma slowly stood up from her chair, looked around, and after gathering herself in the unfamiliar and formidable ambience of the boardroom, said, "Our water problems are very bad, sir. We do want this new plan. But the problem is also with the political leaders; they do not do much with the water problem. They always tie up with the valve man and give us a lot of problem in releasing water. Sir, can you do something about that?" The councilor of Manjunathanagara, who was present during the meeting, looked very unhappy and immediately rose to his defense. He spoke in broken English: "No, sir, I have no connection with the valve man. These slum people have relationships with them and then when things do not work out, they put the blame on me." The IFC staff member, finding herself unable to comprehend the situation, asked, "Why should the councilor interfere? Once BWSSB supplies you with water, you pay for it and that's where it starts and ends. It's between BWSSB, the provider and you, the consumer." She entirely dismissed the fact that the councilor was after all an elected representative and commanded immense power in his constituency.

One of the engineers tried to explain the situation to the IFC member. "They have not paid for water before; this will be the first time. Also the councilor gets his vote based on him continuing with a guarantee of free water. He has a vested interest here." The councilor, reading this as an indication of support for him, chimed in. "I have always given you free water, right, Ratnamma?" Ratnamma, by now quite comfortable with the surroundings, responded, "Yes, that is true, but you have also stopped us from getting water. It's at your disposal."

At this point, another IFC member reasoned, "See, that's why you need to create a market for water, like you have for other things. That way, the councilor will have no role, like he does not play a role in what kind of car you choose to drive. You pay; we supply. It's simple." She turned to the chairman, expecting his approval, but the chairman looked at her askance. From his years in the bureaucracy, he was acutely aware that while India often succumbs to demands of international donor agencies, the demands are far removed from the lived reality. It is also interesting to note that the refusal of the IFC staff member to pay attention to the role of the councilor is symptomatic of a deeper disconnect that plagues development projects across the world. The APSA volunteer translated these words for the Manjunathanagara residents, and a silence descended. They uttered nothing, but a new and more important concern surfaced in their anxious eyes. "How much do we have to pay?" In the meantime the councilor turned to stare at them with annoyance.

Legal Categories

The above meeting is significant not only because the Water Board did not list urban poor as a category in their Act. I was more intrigued by the fact that the poor need to exist as a legal category to avail basic amenities, while there is no parallel discourse regarding the legality of the "middle class." I spent the following weeks talking to various engineers at the Water Board trying to gather their understanding of the urban poor issue. The chief engineer, Basavaraj, was categorical: "You heard what we said the other day, that we have given water to the poor always. It is not a new thing for us. But if the funding of GBWASP depends on putting urban poor in the Act, we will have to do it." "But to even know who the urban poor is," I pointed out, "you will have to do a whole study of the city and submit a report to IFC for them to approve it before it can be included in the Act." "Yes, that's the problem. All they want is people who live in the slums, but what about the floating population who come to the city to construct buildings? They have been receiving water from the public fountains, but if we have to stop that and install metered connections, where will these people go?"

"As far as I gathered from the meeting, you need to form a committee to determine who will be considered as the urban poor. So has a committee been formed?" I inquired. "Yes, you are right," Basavaraj explained. "The committee is not yet formed but internally we have started talking about it. Salma Sadikha will head it; she is the social development officer. She was at the meeting—you must have seen her. Talk to her. She can give you more information on this." Another senior engineer, who would be pivotal in the committee, suggested that I start by looking at the various para-statal agencies who work with and list urban poor as a category in their respective Acts: "That will give you an idea and also possibly these agencies will also be part of the committee to give us some direction. You might want to start with Karnataka Slum Clearance and Improvement Board (KSCIB)."

I spent the next afternoon at the KSCIB. Basavaraj had handed me a report of a study commissioned earlier by the Water Board to understand the urban poor situation in the city. The report had a section on KSCIB:

> The Karnataka Slum Clearance Board (KSCB) has officially recognized 361 slums in the city of Bangalore following a survey they conducted in 1999. The survey lists 103,037 houses with an estimate of 573,556 residents; this comprises about 10% of the city's population. However this estimate does not recognize the unidentified slums, which would then augment the urban poor to 30% of the city's population.[4]

The spatiotemporal organization of public offices such as the Water Board is fluid compared with the strict regimen at Infosys and Janaagraha. When I asked to speak with them, most officials responded by saying, "Why don't you come in some time tomorrow." In contrast to Infosys' time-thrift, time was abundant here. Further, I learned that a conversation about public governance required a different pace, comparatively more helical than the precise steps of coding software.

I spoke to the officer responsible for the BMP area, R. Mahadevan, a Karnataka administrative services (KAS) officer. He handed me a copy of the Act to review their operational definition of urban poor:

> According to Section 3(1) of the Karnataka Slum Areas Improvement and Clearance Act, 1973, Declaration of Slum Areas—
>
> (1) Where the Government is satisfied that;
>
>   a) any area is or is likely to be a source of danger to health, safety or convenience of the public of that area or of its

4. TCE Consulting Engineers Limited, *Preparation of Detailed Project Report for the Work of Providing Individual Water Supply House Connections for the Slums in Bangalore*, submitted to the BWSSB, Cauvery Bhavan, Bangalore, September 2004 (p. 3).

neighborhood, by reason of the area being low-lying, un-
sanitary, squalid, overcrowded or otherwise; or

  b) the buildings in any area, used or intended to be used for
  human habitation are:

  • in any respects unfit for human habitation; or
  • by reason of dilapidation, overcrowding, faulty arrangement and design of such buildings, narrowness or faulty
  arrangement of streets, lack of ventilation, light or sanitation facilities, or any combination of these factors,
  detrimental to safety, health or morals,

  It may, by notification declare such areas to be a slum area.

(2) In determining whether a building is unfit for human habitation, for the purposes of this Act regard shall be had to its
condition in respect of the following matters, that is to say—

  • repair
  • stability
  • freedom from damp
  • natural light and air
  • water supply
  • drainage and sanitary conveniences
  • facilities for shortage, preparation and cooking of food and
  for the disposal of waste water, and the building shall be
  deemed to be unfit as aforesaid, if it is so defective in one
  or more of the said matters that it is not reasonably suitable
  for occupation.

The Act has a distinct territorial component: the "slum," which is
considered the usual habitat of the urban poor. However it does not offer
much information about the population as such their means of livelihood,
income, health, and so on. The following is a conversation between me and
Mahadevan.

MAHADEVAN: Our task is to classify the slums in the city and work
    towards improving them.
SDG: But isn't your agency also about clearing slums?
MAHADEVAN: See, that's the important part. Clearance and improve-
    ment are in a way linked. Improving the living conditions in a
    slum will change it into an area that will no longer look like a
    slum, but, say, a lower-middle-class neighborhood.
SDG: Then what about the reports of slum evictions that we see in the
    newspapers?
MAHADEVAN: That we have to do sometimes because as the Act says,
    if the area is absolutely dangerous for people to live, we have to

evict them. But that is a rare thing. We try to do some work and better their condition.

SDG: Is there anything specific you do about water supply?

MAHADEVAN: Yes, we give legal status to the slum by declaring it and then the BWSSB supplies them water and as you know some slums in Bangalore already have metered connections.

SDG: So there is a difference between "recognized" and "declared" slums?

MAHADEVAN: Yes, all slums are recognized, but not all are declared. Declared means we have to be sure that the land it legally owned by the slum dwellers or donated by some agency or given to them by the government.

SDG: And land title also ensures they have water supply.

MAHADEVAN: But none of the basic amenities like water can be given on disputed land.

SDG: There is, however, a legal provision where BWSSB does supply water to the slums even when they are on illegal land, but there is a legal document one has to sign where BWSSB will discontinue the connection without getting into the hassles.

MAHADEVAN: Yes, that's right, but BWSSB does supply free water through public fountains where the slum dwellers get water as well as the pavement-dwellers.

Once again, it was clear that the urban poor issue cannot be contained merely within the slum; the population that lives on the many streets of Bangalore has to be taken into account.

Following this, I visited the DMA, one of the stakeholders of the Water Project, responsible for supervising the CMCs. There I met Varamballi, a retired bank officer who now works with DMA for Swarna Jayanti Shahari Rojgar Yojana (SJSRY). This program, Varamballi explained to me, was exclusively designed for the urban poor to provide training in various skills, such as plumbing, technical repairs, sewing, and so on, "to make them eligible for employment or start their own business." Under SJSRY, DMA identified the poor households where the income is equal to or below Rs. 23,500 rupees per annum (approximately three hundred and seventy dollars). "Then, can the SJSRY, which is an existing project, help identify the urban poor for GBWASP?" I wanted to know. Varamballi responded, "It could, but the urban poor is a very tricky category. Who can say for sure that if I live in a slum, I am an urban poor? They can be under government schemes which they benefit from, and in the kinds of jobs they do, there is no way of telling how much money they earn; it's all very hazy."

Varamballi's exposition of the urban poor opened up yet another issue, the politics of poverty and the fluidity of the informal economy. This also points to an ambiguity intrinsic to the urban form: While there is the rhetoric of eradication of poverty, poverty is a collateral of the structures of social inequality that sustain the city. Sassen provides part of the answer. In her analysis, the global city "is a strategic site for disempowered actors because it enables them to gain presence, to emerge as subjects, even when they do not gain direct power" (1998: xxi). The argument is compelling in the case of Bangalore, an aspiring global city. The poorer population is readily available for unskilled jobs as construction laborers, domestic help, janitors, chauffeurs, and so on.

The Place of Water

Typically, water in India straddles two kinds of narratives: symbolic-spiritual and material-juridical. The Water Board building is named Cauvery Bhavan because its main source of water is the river Cauvery,[5] a perennial river about 100 kilometers (approximately 60 miles) from the city of Bangalore. BWSSB also draws water from three other rivers, Arkavathi, Hemavathy, and Himsha,[6] but at a relatively lower volume. Originally Kaveri and then said to be renamed Cauvery by the British, it is one of the *Saptasindhu* or the "Seven Sacred Rivers" in the Hindu mythology. The symbolic importance of the river is captured in the Water Board logo and a stone statue of Goddess Cauvery (pouring water from a pitcher) within the Water Board premises. Further, the Water Board website at that time played a video of "Talacauvery," the origin of Cauvery in the Coorg region, a hilly plateau bounded on the west and south by the Western Ghats in Karnataka. Talacauvery has deep religious significance. Legend has it that Goddess Cauvery makes her appearance once a year during Tulamasa, when thousands gather to take a sacred dip. The video displayed a streaming script: "Born of the heavens, Embraced by Mother Nature, Entrusted to us to preserve for future generations, Water for life." Further, water is symbolically thought of as sustaining and life-giving in everyday life.

Materially and juridically, I often encountered various questions raised about the place of water within the Indian Constitution: Is water a fundamental right? Who is responsible for supplying water—central, state, or local government? Is there any constitutional provision for free water especially

5. Cauvery is also disputed water resource between the neighboring states of Karnataka and Tamil Nadu.

6. "Keeping in view the long range requirements of the city and dependability of the supplies, the Committee recommended to consider tapping the perennial river Cauvery rather than any of the other three sources. This recommendation was accepted by the Government during April 1964 and administrative approval was accorded to the Cauvery Water Supply Scheme (CWSS) 1st Stage Project. Work with an estimated cost of Rs. 22 crores was started during 1969. Construction of this project was completed in about five years and the supply of Cauvery water to Bangalore commenced from 24th January 1974" (available at http://www.bwssb.org/water_source_schemes_cau very.html [accessed September 12, 2007]).

to the urban poor? How is the water revenue structured? These questions surfaced and were important because the constitutional placement of water is surprisingly rather complicated. If one were to follow the three Constitution lists—the Union List, the State List, and the Concurrent List—water does not appear in the Concurrent List but appears in both the Union List and the State List. The State List, entry 17, is as follows:

> Water, that is to say water supplies, irrigation and canals, drainage and embankments, water storage and water power subject to the provisions of Entry 56 of List 1.

Entry 56 of List 1 in the Union List states the following:

> Regulation and development of inter-State rivers and river valleys to the extent to which such regulation and development under the control of the nation is declared by Parliament by law to be expedient in the public interest.

Ramaswamy Iyer (2003) argues that the simultaneous inclusion of water in both the lists is a source of interstate conflict mostly regarding the sharing of water from rivers than run through several states. The dispute of sharing water from the Cauvery between the states of Karnataka and Tamil Nadu is particularly relevant. The Cauvery dispute, however, was seldom a topic of discussion in the course of my work at the Water Board. As one officer explained to me, "The negotiations are usually carried at the Center (Central Government in New Delhi) and at the political level; we at BWSSB hear of the decisions only as a para-statal agency later."

The emerging prospect of paying for water further complicated the constitutional ambiguity in relation to the urban poor. Slum residents often discussed among themselves whether the Water Board had permission to charge them for water from Delhi, meaning the Central Government. These doubts spun an affective discourse that combined the sanctity of water with social justice toward the urban poor. Interestingly, as Iyer contends, the Constitution does not include "any *overt* reference to water as a basic essential for life, and therefore a basic human and animal right" (2003: 24; original emphasis). However, he notes that natural right to water has subsequently emerged as a legal extension of the fundamental right to life as a result of various legal cases and acts.

Still, the ethnographic data on free water is disconnected from these acts. It exists as an affective and a class discourse of the underprivileged; right to water is considered indelibly scripted in the Constitution at its inception. "It cannot be otherwise," opined one of my informants. "The government knew from the very beginning that water is sacred and to charge poor people like us for water is simply not right. In this country we never deny water to the thirsty and the dying!" Hugh Raffles, drawing on Gaston Bachelard's formulation of the relation between human beings and water, writes: "The language of water is not metaphorical . . . the language of water is a direct poetic reality . . . in short, a continuity between the speech of water and the speech of man" (2007: 314). The spiritual continuity between Cauvery and the sanctity of water in people's imagination overlaps with Bachelard's understanding. This continuity is, however, oblivious to the uncertain place of water as a legal category. When I mentioned this uncertainty during a conversation with some residents in Manjunathanagara, it was vehemently disputed. Further, the omission "if at all was true" was considered by them as demeaning to human life: "Why would one need laws for giving water? It's common sense! We always give water to the thirsty and the dying; it is a sin not to do so."

The spiritual discourse strongly resonated at the Water Board, as well, especially among engineers who worked directly with the urban poor, installing connections and conducting maintenance work. I often came across observations such as the "BWSSB has always been supplying water to the urban poor free and it's a good thing to do." In other words, while they were relatively convinced that privatization would earn the Water Board much-needed revenue, the change was problematic ethically and spiritually.

Water Market Meetings

Salma Sadikha, the social development officer of BWSSB, was present at all the meetings regarding the urban poor issue. Her new task focused on the pro-poor program where water would be easily accessible to the urban poor while the economic interests of the Water Board would be preserved in its changing role as a provider. I met Sadikha one afternoon in her office on the seventh floor of the Water Board building. "I am interested in the idea of privatization of water," I said in response to her query about my interest in the Water Project. "Why water and why urban poor?" she wanted to know. I explained, "There is a clear initiative at BWSSB to create a water market, that is, shift water from a basic amenity to a commodity." Overall, I suggested that though water continues to be a basic amenity, the basis of water supply is moving in a new direction, and I am interested in how the poor are coping with this change.

> SDG: I am also working with Janaagraha, but there seems to be some hesitancy on their part to include the poor as full participants in the project.
> SADIKHA: That's because they never try to know. How much time do they spend at the slums anyway? Yes, you are correct, BWSSB wants to change some ways in which we supply water. It's easy

for the middle class, but it is a different issue with the urban poor
because they have really never paid for water.

SDG: Are there any specific measures you are taking toward this?

SADIKHA: I work on bringing in two kinds of reforms, finding different ways of connections to lessen the fees, and relaxing the land
tenure requirement for installing piped connections, which is a
problem for the urban poor's access to safe drinking water.

Sadikha reiterated the connection between land rights and water supply that I discussed earlier. However, she drew my attention to the various
methods by which the urban poor can connect with the Water Board supply:

SADIKHA: I offer them two alternatives, individual and shared. The
individual connection is one that is the general practice, but
shared connections can be arranged among various families who
agree to split the bill among themselves; this seems to be a good
alternative for the urban poor.

SDG: How are such arrangements made among families? What are the
details of the arrangement?

SADIKHA: I tell the slum dwellers about the kind of connection we
offer. It's up to them to decide after the meetings.

SDG: Which meetings are you referring to?

SADIKHA: I hold regular meetings in the slums before any water connection is approved. Well, I am holding one next week, and you
are welcome to attend it.

On the day of the meeting, Sadikha picked me up from the bus stop we
had previously decided on. As her car (provided by the Water Board) neared
Nellorepuram, the road narrowed, finally leaving the driver no room to pro-
ceed. We disembarked and walked along the winding path that finally led to a
small convenience store at the entrance of the slum. A group of men were assembled in front of the store, and one of them came forward to greet Sadikha.
He introduced himself as Venki and "the leader of this place." "This way, please,"
he said and gestured for us to move to our right. As we entered the slum, in a
loud voice he directed the people around us: "Go to the school building, now."

The school comprised two rooms and was part of a public program named
Anganwadi (courtyard shelter), designed to offer basic education to preschoolers
along with lunch as part of the Mid-Day Meal Scheme (MDMS). *Anganwadi*
was designed to keep children in school because they often dropped out to work
in order to supplement low family income. The *Anganwadi* also served as a day
care center since most of the parents, especially the mothers of the toddlers,
were at work, mostly as domestic help, during the morning hours. Looking
around, I was curious about the adjacent room, which was dark but had signs

of activity. I peeked; the low flame of a kerosene stove illuminated the room faintly. In a large aluminum cooking pot rice, lentils, and some vegetables were being cooked. One of the two women present in the dark room, who was the *Anganwadi* teacher, said, "The light bulb is out, so we have to cook in the dark. I have asked the councilor to replace it, but who knows when he will." Perhaps gauging my concerned look, she smiled at me and said, "I know they can, but here we have very few things to work with. I think it is more important that they have the food rather than not because then their mothers will blame me. The food is more important; they come here mostly to eat than to read."

I slowly walked back to my chair. By then a motley crowd of women, men, and children had gathered for the meeting. Sadikha introduced herself as a "BWSSB officer" designated for "social work in the slums of Bangalore." "What is your current situation with the water supply?" she wanted to know. While the women in the room remained quiet, exchanging glances with one another, several men raised their hands. Sadikha invited one of them to speak. "We get water two to three times a week, but then the time is not fixed, so we have to be ready. It can also be very early in the morning." The erratic supply of water, especially at inconvenient times of the day, was a major complaint in most slums. Besides the overall scarcity of water, part of it, as I mentioned above, relates to the personal disposition of the valve man and his tenuous relationship with the residents.

Sadikha offered the solution of multiple connections that BWSSB would like to promote. "In this plan" she explained, "two to three families will connect to one connection, and the monthly bill will be split equally among them." Sadikha also assured everyone that BWSSB had arrived at this decision after careful deliberation based on several reports prepared by international donor agencies, such as the Asia Development Bank, AusAid, and Water and Sanitation Program, among others. The reports encouraged multiple connections mainly to address and stem the problem of illegal connections, which Sadikha, for obvious reasons, did not mention.

While illegal connections are generally conceived of as a practice prevalent among the poorer sections of the city, extension pipes jutting out from houses in the middle-class neighborhoods I frequented were not a very uncommon sight. Some middle-class citizens explained that the delay in the bureaucratic process forced them to circumvent the system and pay bribes to install a connection illegally. Yet illegality is normatively associated with the urban poor. On the other hand, the middle class is exempted for compromising its ethics when confronted with a corrupt bureaucracy.

Technically, illegal connections lead to what is known as "unaccounted for water" (UfW), resulting in significant loss in revenue for the Water Board. The World Bank defines UfW as the "difference between the quantity of water supplied to a city's network and the metered quantity of water used by the customers. UfW has two components: (a) physical losses due to leakage from pipes, and (b) administrative losses due to illegal connections and under registration of water meters. While every case is different, often both components contribute roughly equally to UfW."[7]

It soon became clear during my conversations with the Water Board officials and WSP staff that, besides economic loss, UfW was also seen as a vital issue of illegality that needed to be resolved urgently. The urgency nonetheless exclusively focused on the urban poor usage or, rather, misusage of water. Often when I broached the issue of UfW in middle-class neighborhoods, it was casually dismissed by both the Water Board and WSP as a necessity in the face of government corruption. In other words, UfW was an ethical frontier that distinguished the middle class from the urban poor. Further, the prevention of UfW was also necessary as the raison d'être to do justice to the "law-abiding" middle-class citizens vis-à-vis the "delinquency" of the urban poor and the state. The water market in turn would remove the bureaucratic constrains as well as the possibility for corruption since performance of the provider would be the sole determinant for the service to continue. The WSP in particular was very vocal about the menace of UfW and the concomitant solution of bringing the urban poor within the folds of the water market through private-sector participation. The WSP team handed me a few of their reports that they suggested would help me better understand the relationship between poverty and water procurement. One such report, titled "Serving Poor Consumers in South Asian Cities," had a section on the "Nature of Poor Consumers," which is as follows:

> Service providers . . . often resist serving the poor and sometimes simply ignore them. . . . Firstly, low income consumers are often perceived to be high risk, low return customers—the cost of serving them is high. . . . Poor communities often present a challenge because they live in areas which are dense, with unplanned layouts and narrow roads and often on land which is insecure. . . . A formal contractual relationship directly with a client household is considered impossible. . . . Secondly, many governments establish policies which explicitly prevent "informal" settlements from accessing all but the most basic municipal services until the land on which they reside is "regularized" in some way . . . the objective of any reform, including the introduction of the private sector, must aim to reduce the "distance" between the utility and its poor customers.[8]

---

7. Available at http://web.worldbank.org/WBSITE/EXTERNAL/COUNTRIES/MENAEXT/EXTMNAREGTOPWATSUPSAN/0,,contentMDK:20537897~isCURL:Y~pagePK:34004173~piPK:34003707~theSitePK:586013,00.html (accessed May 21, 2008).

8. "Serving Poor Consumers in South Asian Cities: Private Sector Participation in Water and Sanitation," available at http://siteresources.worldbank.org/INTPSIA/Resources/490023-1120845825946/sa_psp_sa.pdf (accessed October 8, 2007).

Sadikha's work, which offered a direct link between the urban poor and the Water Board, was crucial to the WSP team to initiate the idea of the water market in the slums. Often either Sadikha would accompany the WSP team or appoint a guide to go on what were generally known as "slum visits." The WSP team would collect "firsthand experience of the scarcity of water" on these visits. On some occasions, Sadikha requested me to accompany the WSP team. By that time I was well aware that such slum visits were routine and were usually scheduled to "document the dire situation of water supply and to justify the need for a new project." It is interesting that none of these visits were scheduled in the middle-class neighborhoods because water privatization was not considered to be an issue in these areas. WSP also explained that with the "educated middle class the question is about good service, not how much needs to be paid. They have been paying for a while but for no service at all. This is where we come in."

The task of WSP team members was twofold: First, they had to assure the poor of transparency, that is, unlike the state the private provider would not supply water in exchange for votes. Second, accountability, that is, privatization, would be a "good thing for them" since the provider would be held to a rigorous standard of delivery, and failure to do so would result in dismissal. Like other developmental agencies, WSP regarded water and water supply as an amenity and a technical matter, respectively. Additionally, given the "best practices" developed by international donors and organizations, WSP too relied on the modular quality of the practices, which were transportable and implantable across the globe. Thus the slums were seen as the homogeneous space occupied by the poor and not as a meaningful habitat were lived experiences mattered. But to think of WSP and similar organizations as blatantly ignorant is to miss the deeper issue. These organizations have changed their rhetoric to address criticisms and reestablish their power as more "participatory." The emergence and establishment of international donor agencies to redefine not just resource allocation but also nature itself is important here. Michael Goldman's *Imperial Nature* offers insights that are broad and also specific to the experience of privatizing water. Specifically Goldman's argument that developing agencies now seamlessly dissolve and contain the apparent tension between neoliberalism and social justice has led to the emergence of "green neoliberalism." Green neoliberalism, on the other hand, has led to transformation of the World Bank (and other development agencies) in ways that it has to now mandate visiting previously peripheral sites, even if briefly, such as slums.

I offer the following vignette as a paradigmatic example of a "slum visit." in Manjunathanagara. After a long arcane speech by WSP about the value and convenience of privatized water, the residents gently but categorically turned down the offer. This would not halt the Water Project by any means, but this ethnographic piece is a window into the interstices of power. The residents, all men, who spoke did not offer any specific explanation for their decision but simply stated that they were not interested in having "water just outside our doors." The WSP team was taken by surprise; one of the members attempted another round of convincing. "You will have water, right *here,*" the member said, pointing to the wall next to one of the dwellings. The residents, including the women this time, waved their palms side to side and nodded to once again convey their definite no. As we got into the car, one of the WSP team members said in exasperation, "Now, who does not want water? This is completely . . ." He failed to come up with an appropriate adjective. Later on when I discussed this with the local councilor, who was not present at the meeting, he seemed content with the response of his electorate. "Madam, they are not used to the rich [meaning WSP] people; they are outsiders. I am an insider, one of them, and I always help with whatever they need." The WSP team overlooked the fact that water as a mediated resource is entrenched in politics, where the local bearings cannot be easily dislodged. In his work on water supply to "settlers" in Mumbai, Nikhil Anand (2011) has drawn attention to the nuanced politics of access and distribution. Anand writes:

> Through the process of incremental and differentiated citizenship, qualified settlers are required to substantiate applications for water connections with a formidable set of documents to prove their resi-dence. Because this coterie of documents . . . is difficult to assemble, settlers do not approach the water office directly but instead approach city representatives for "help" in getting water connections. As they mobilize diverse relations with democratically elected representatives, dadas, and social workers to access water, settlers constitute them-selves as citizens and subjects in the city. (2011: 546)

Normalizing the Market

As I discussed in the introduction, the constitutive nature of the market marks a departure from classical liberalism to contemporary neoliberalism. In other words, the move is toward normalizing the market as though it were a "natural" tool to augment efficiency and arrest corruption. Further, the state was expected to embrace and emulate the market as the only redemptive tool available. The normalization was visible in the water meetings that I attended in the middle-class neighborhoods; the impending rise in the cost of water was seldom questioned. The majority of my informants at Infosys, many of whom also lived in the Greater Bangalore area, were not even aware of the Water Project, since water was not an issue in their everyday lives, whether at work or home. Further, when I introduced them to the Water Project, most of them upheld the privatization of water as a moment of arrival for the Indian polity.

Besides economics, the middle-class normalization of the market entailed a certain understanding of what it means to live a good life, that is, the sense of well-being. It released life from the drudgery of struggling for something as basic as water. "I remember," said one of the attendees in at the Pai Layout RWA meeting, "the line of colored buckets in my bathroom filled with water because we would have to go without running water for the next few days. We would have to keep a close watch on all the members of the family so that nobody would use more than needed." The increased payment was considered a magic turn where one would have a 24/7 water supply. However, whether BWSSB would be able to supply water 24/7 also remained unscrutinized and unexplained. On the other hand, one cannot deny the difficulty that plagues everyday life in the face of water scarcity. However, it is problematic when the end of this scarcity is cast in terms of the market and class privilege that ignores those who are unable to participate in it.

In analyzing the market, Gerald Berthoud argued that "to be human is to strive to escape from constraints, both natural and social, to become an independent individual . . . being human means being independent through the use of limitless technological innovations and a boundless market" (1991: 82). The discourse of scarce and impure water is prominent across the developing world. It reflects a poor standard of living and the meager value placed on human life in general. Bottled mineral water, for instance, in India, is used mainly for reasons of hygiene rather than convenience or "surplus health." A failing and corrupt state is perceived to be responsible for this demeaning status for the polity both for itself and the outside world. Hence the introduction of the market would counter not only the failure of the state but the omnipresence of potable water, even as a conjecture will raise the quality of life and enhance India's standing in the world. In this sense water, like software, becomes a parallel venue to capture how the market changes the imagination of the Indian polity.

There was a mutual silence about the increase in the price of water between the middle class and the Water Board. The understanding was, as I gathered from my conversations at the RWA meetings, that privatization, based on the "accountability" of the operator, will "obviously" ensure a reliable and uninterrupted supply of water. Looking at it from the supply side, there were several issues that were ignored in this expectation. For instance, topographically Bangalore is a plateau, and water, which travels a considerable distance from the main source, the Cauvery, then has to be pumped up the elevation. The engineers at the Water Board often discussed how the topography of the city was a challenge for the desired 24/7 water supply. The market narrative of consumer and provider, accountability, and transparency obscured the actual reality of water sourcing and distribution. The market also made the poor increasingly invisible to the middle class, who was now once again the vanguard of emerging India. Finally, when the state is gradually removing itself from welfare programs to yield way to the market, it desensitizes the middle class further to the dire situation of the poor.

While words like "privatization," "market," "accountability," "transparency," and so on, were common at Infosys, Janaagraha, and RWA meetings, these words were unknown to my informants in Manjunathanagara. Nevertheless, the critique of the market emerged primarily among the poor as a question of survival and livelihood. Broadly, the privatization of basic amenities, especially water, had led to a slum movement in the city of Bangalore. At the time of my fieldwork, the movement did not have a strong base in the slums of Greater Bangalore, which were still peri-urban, and the residents were quite not yet familiar with the specific nature of urban politics, especially those pertaining to urban renewal. To give an idea of the movement, I will briefly discuss the activism of Issac Arul Selva, a slum resident himself in Lakshman Rao Nagar. Central to the movement is a newsletter titled *Slum Jagattu* (World of the slums), which Selva started in 2000 to voice the concerns of slum dwellers through their own writings.

When I met Selva in 2006, it was clear that he politically claimed his subject position as a slum resident and engaged from that position of underprivilege. He overtly stated that their collective condition of underprivilege was "unrepresentable" and likewise he was also somewhat reluctant to talk to somebody like me with an upper-class-caste background. He argued that outside representations by "someone like you could not and will not help overcome" the class and caste (given that most dwellers in his slum were also Dalits) divisions. He gave me a copy of the latest issue of his newsletter and pointed to the stories of everyday struggles that are the reality in the slums. He followed up by saying, "These are in the words of those who live here," carefully emphasizing "here" to indicate the territorial importance of the slum for these experiences. *Slum Jagattu* was widely circulated and read by people residing in Bangalore slums, and it had also become a platform of resisting neoliberal urban plans that tend to disproportionately dislocate the poor for megacity projects. Drawing on his experience as a slum dweller and that of editing *Slum Jagattu*, Selva had constructed a website on developmental issues, where water features prominently.

Selva's resistance to neoliberal changes in relation to water, offers us a useful way to think about "structures of feeling" proposed by Raymond Williams in his discussion of how the narrative of the social is often the narrative of the past: the social as formed (Williams 1977). Williams argues, "Social forms are then often admitted for generalities but debarred, contemptuously from any possible relevance this immediate and actual significance of being" (1977: 130). He insists on feelings rather than "ideology" and "worldview" because "we are concerned with meanings and values as they are actively lived and felt" (132). The "structures of feeling" in the instance of multifarious resistance are not recognized or narrated as formations of the past but emanate in the present as slum residents negotiate the impending water market.

With Sadikha's visits it was gradually becoming apparent to the urban poor that now there would be a formalized way to access water. It was perceived as a move away from an unstructured and collective method of gathering water at the impulse of the valve man or the benevolence of the councilor to a more structured individualized system. On the other hand, as an employee of the Water Board and for her dedication to the poor, Sadikha continued to embody the state as a parental figure, the *mai-baap*, a relationship that historically configured the interaction between the state and the poor. Sadikha also offered them ways in which the Water Project would lead them on a path to full citizenship. For example, Sadikha assured them that the monthly bill from the Water Board could also serve as a proof of residence for other public and private services, such as electricity or purchasing property.

As my ethnography in Manjunathanagara progressed, it became increasingly evident that the unease with payment for water was deeper than it first appeared. For a population that had received free water, this was definitely a change, but the real unease with the new plan lay elsewhere. The urban poor have been paying for water they needed beyond that supplied free by the state; they would typically buy water from private operators who supplied water in tanks. Often residents rhetorically asked, "Why is the government doing this to us?" The standard response would have been that the new network would reduce the quantity of UfW, which is disrupting the efficiency of the Water Board. However, like me, they also understood the middle-class politics that was driving the privatization initiative.

Over our conversations, it emerged that many were deeply concerned about the presence of WSP and IFC members in the work of the Water Board. They especially perceived their "slum visits" to "assess the situation" as transitory, detached, and spasmodic. It was not that the residents, as they themselves mentioned, were unaccustomed to these kinds of visits organized by city NGOs. "This is different," as one resident said. "They are coming here with the government." The direct collaboration between the state and the international donor agencies in the Water Project instilled a different kind of uncertainty among the residents. It signaled the withdrawal of the state. It was also clear to them that a lot of changes the Water Board was adopting were under the direction of WSP and IFC, and one could never be sure how far this contract between the two would affect the existing welfare schemes for the poor.

It was standing joke among the residents in Manjunathanagara that they were "really good" at community participation—a practice that was very familiar to them—"because we are always being asked to do so." The mockery in their tone signaled their sense of deep distrust of the entire participatory paradigm as a way to alleviate poverty and institute democratic ideals. One late afternoon when the day's chores were done, I sat in the shade of a tree inside Manjunathanagara with a group of young men and women discussing the water issue. "Very little changes for us," observed one of them. "We know that. We 'participate' because sometimes we get paid and sometimes we do them a favor because they need poor people like us; that's what the foreigners [donor agency members] want them to do." Another person offered a rejoinder, "It also gives us an opportunity to do something else; sometimes the NGO ladies make a lot of sense when they talk but in the end nothing really changes." "The NGOs do some work for us but the real work, the government does for us, but if now we have to pay for water, then we have to be worried. Why is the government listening to foreigners? Because they want their money?" "Then at the meetings (of political parties) they will talk about how bad the British was and how Gandhi fought them to be free," came another rejoinder. "Then what's the use if we are again going to ask foreigners to help us? They will again start ruling us as they did during the bad days."

While most slum residents do not technically use water free of cost, as there are other costs involved, the role of the state, as the *mai-baap*, the benevolent parental provider, was nonetheless reassuring. Further, in these conversations, linking the state with the market in the sense that there has to be a fee paid to secure water violates a long-established regime of paternalism, and the presence of "foreigners" escalates the shifting relation at the level of suspicion and correctly so. In another sense, the suspicion takes the shape of the fear of yet another form of colonization. The market in reality is undoubtedly a form of colonialism for it territorially constricts the polity, limits ideology, and constrains practices solely through the idiom of money. While it was shrouded in silence in the middle-class conversations, the urban poor openly vocalized their concerns about the market, though the word itself was never uttered. Instead, the residents used "payment" as a shared vocabulary to comprehend the actual changes that would soon affect their lives. The difference in the class reaction provides powerful insight into how the apparent neutrality of the market is after all a fiction.

The Issue of Integration

The Water Project team at Janaagraha was quick to perceive my interest in the urban poor, and I was assigned the responsibility of documenting the meetings that focused on the issue. This was not intended to be a record of the meeting minutes, which was already a customary practice, but a detailed documentation of the debates through which Janaagraha would arrive at their stance toward the urban poor problem. Ramanathan encouraged me to "document the *process* through which we will arrive at an ideology of how we feel the urban poor should be addressed. This will also provide a blueprint for other projects that we will undertake in the future regarding urban poverty."

I present the following document, which I was asked to title "Issues in Focus," to highlight the varying arguments that were offered around the urban poor theme. The idea was to capture the complexity of the theme and the diverse positions taken by various team members, and above all to underscore the class politics through which the urban poor was being constructed.

Issues in focus:
1. Is Janaagraha endorsing the beneficiary capital contribution proposed for the poor?
2. Is Janaagraha endorsing the tariff structure for service delivery?
3. How does Janaagraha facilitate the participation of the urban poor?
4. What is Janaagraha's proposal for water supply in the slum areas?
5. How does Janaagraha address the above issues in the context of the existing community facilities?
6. How does Janaagraha address the issue of the floating population (those outside the slums)?

Discussions on May 9, 2005
MEMBER A:[9] The problem with the urban poor issue is in the way Janaagraha has termed it, that is, "integration." It is left too loose and is now open to varied interpretations by diverse agencies.
RAMANATHAN: The only way to clarify is to make it crystal clear in the terms of reference [ToR] that Janaagraha is only focused on including the urban poor representative in the Ward Project Committee. We need to emphasize the process, not the outcome.
MEMBER B: Based upon the opinion of other experts in the domain of urban poor, like Salma Sadikha of BWSSB and Renu Mukund, who was part of the team that originally designed the GBWASP [Water Project], the only way one can be assured of integration to some extent is to work in the slums, i.e., at the grassroots level.
RAMANATHAN: Janaagraha cannot bear the sword of Damocles to ensure urban poor participation. We will provide space for the poor to participate and since other NGOs are working with the poor, there is little need to reinvent the wheel.
MEMBER A: In that case we need to connect to the NGOs.
RAMANATHAN: Let the NGOs bring the poor on to the Ward Project Committee [WPC].

---

9. All names have been removed, except Ramanathan.

MEMBER A: With the middle-class areas the design is through; with the poor one needs a different approach" and I do not know who is doing it.

RAMANATHAN: Janaagraha's role needs to be made clear in the ToR where "integration" is defined as a space which is made available for the poor to participate while at the same time not assuming responsibility to guarantee that the poor will attend the WPC meetings. Let all the NGOs working at the grassroots level coalesce around this project so at the end Janaagraha is not the only one with the onus of making this a successful project.

MEMBER B: With the current ToR that is being prepared with USAID I will highlight that "integration" of the urban poor will only be through the urban poor representative at the WPC level and not at the grassroots.

MEMBER A: There are two related issues in this: one is about the sanctity of the project and the other is the economics.

RAMANATHAN: This may well be an ideological divide that needs to be debated and discussed, and Janaagraha will be forced to take a stand, but there definitely needs to be a discussion amongst ourselves before we arrive at it.

MEMBER A: This brings to focus that the urban poor situation is different from the middle class also in that the latter have a set of choices of the kind of water connection they want, whether shared or individual. How does the community decide and what role does Janaagraha play in helping them make the decision?

RAMANATHAN: It will only be possible through the WPC where if the urban poor come to the meetings, they will have a fair space for discussion, but only through their representatives.

MEMBER A: How are the poor going to be educated in the choices they have?

MEMBER C: The training manual that I am doing for the WPC members will carry a set of best practices implemented by agencies like the WSP especially in regard to the slums and the poor.

RAMANATHAN: If we continue to think that the poor cannot help themselves, we are wrong. For choices, there are local mechanisms in the "bazaar of choices" and the poorer population will know what is best for them.

MEMBER A: But the messaging for the poor has to be significantly different since their context is different.

RAMANATHAN: It is not "different" but "incremental" in that the urban poor representative admittedly has a more difficult job to perform.

MEMBER A: The issue is also with the process of decision making and Janaagraha does not seem to be doing anything in this area since we are not training the citizens in this respect.

RAMANATHAN: We cannot take on the responsibility of ensuring that the urban poor go through a process of making choices and then emerging out of a process; all we can do is provide a space in the WPC for him/her to participate.

The meeting concluded with the following resolutions for the team to work on:

1. Define "integration" of the urban poor in the ToR: Integration to mean that Janaagraha will integrate the urban poor to provide a space of participation along with other members in the Ward Project Committee and will not be involved in any form of grassroots mobilization in this regard.
2. Work on the WSP/WaterAid documents on their experiences and proposals of services to the urban poor to be included in the training manual as "case studies."

Urban Poor and Participation

At a subsequent meeting, one of the WSP staff handed me several "case studies" from work it had done with the urban poor in creating a water market in countries like the Philippines and Brazil, among others. I was responsible for perusing and presenting these documents to Janaagraha. I had to specifically sift the case studies for "participation," which Janaagraha could adopt as a model. However, participation was missing from the case studies that the WSP staff provided. In a context where participation of the poor was being publicized as the path to institute the market democratically, its significant absence in the WSP documents confused Janaagraha. Given the extent to which donor agencies have been promoting participation as the democratic tool in development project, I too was somewhat surprised. In the meantime, WSP continued insisting that Janaagraha actively enroll the urban poor in its participatory model if Janaagraha was expecting World Bank funds. Janaagraha urged me to "look in all possible places to get hold of at least one case study that documents urban poor participation." I discussed this with WSP members, but nobody could successfully point me to a case study that explicitly talks about participation of the poor.

Most case studies included an in-box with a picture, usually of a woman securing water at a faucet with a comment about the difference the new system had brought in her everyday life. Rarely were there any direct quotes included from the citizens whose lives supposedly changed. While one does not expect WSP documents to be ideologically informed, nonetheless there was an explicit mention of how governments in the developing world are opposed to market reforms and how electoral politics can mar the promise of the market. However, the case studies mentioned how various states realized that the market, being "neutral" and "apolitical" could indomitably serve as a panacea to the political "disturbances" affecting water supply. It is also interesting to note that when I wanted to know how they address the issue of water privatization and the urban poor in the light of the Cochabamba Water Wars in Bolivia in 1999–2000, they dismissed the event as an aberration. Therefore, it was not surprising that the case studies had a consistent narrative structure that diligently depicted pictures of the "happy" poor availing "privatized" water, but their participation, or lack thereof, was only a matter of teleology to advance the market.

For Janaagraha it was critical to showcase a situation where consumers, particularly the urban poor, were being *made.* The practice of making consumers in this sense involves people's direct involvement with the market where they make "rational choices" from an array of alternatives available to them. Therefore, if citizens in other places chose a privatized steady system over an unstable public one, they supposedly arrived at their choice only after weighing the pros and cons of the two systems. Citizen participation, as we now know, has become a key organizing idiom in developing countries. As a democratic ideal it has been relocated to civil society after its perceived failure as the foundation of egalitarianism. This relocation is a way to reclaim *lost* citizenship, on one hand, and, on the other, to transform citizenship as an active space to introduce and coalesce market reforms in public governance. Structurally, participation continues to frame the relationship between the state and the civil society, but it seeks a shift in the balance of power to the civil society only to further privilege the already privileged.

In her study of water privatization in sub-Saharan Africa, Sylvy Jaglin (2002) has made a remarkable connection between urban poor participation and water supply in cities. She writes:

> Closely related to the process of building economically viable water services, participation is invoked above all to circumvent two major difficulties, namely assessing demand from the poor and managing systems intended for unprofitable customers. In urban areas where there is mass poverty, a substantial deficit in infrastructure and a largely informal economy, participation of the poor seems to reflect a compromise between the ambition to provide universal access to water and the principle of cost recovery . . . participation is not a miracle solution: there is a considerable risk that the systems it

produces, lacking stability and often resulting in inequalities, will lock particular districts or settlements within the urban areas into substandard systems of service provision. (2002: 232)

I inquired several times at Janaagraha, "What are you expecting the urban poor to participate in?" The urban poor, they responded, need to be empowered to hold the state accountable for its services. Though the relation between the poor and the state, as we saw earlier, is usually one of parental nurturance and benevolence, it was seen as undemocratic and irrational. The resonance between the values of Infosys and its real application by Janaagraha in the social realm was distinctive. "They participate," as one volunteer answered, to "take on a system that has neglected them, and as an educated middle-class member, it is my responsibility that they know their rights as a citizen." "Citizen or consumer?" I wanted to know. "It's both, unless you think like a consumer you cannot rise for your rights." "But during the nationalist movement," I pursued further, "people fought for their right to an independent nation as citizens, not as consumers, and were successful." "It was a different fight. We were fighting foreigners; here we are fighting our own people. It is sad, but now everybody understands money more than anything else. Otherwise we would not have this high level of corruption!"

The democratic implication of participation of the urban poor was too erratic and too convoluted to sort out at once. It began with my search for case studies to gauge the WSP narrative about the "eagerness" of the poor to participate in the water market. Yet participation was rather difficult to come by. The only thread that seemed to give some sense of cohesion to the idea of participation in the Water Project was the financial commitment across classes in the form of the beneficiary contribution. On the other hand, the unproblematic transfer of "accountability" as a value from market economics to the social realm, belies an understanding of power among the state, the NGOs, the international donor agencies, and the urban poor. In a milieu where the urban poor were even reluctant to attend citizens' meetings organized by Janaagraha and Infosys and intimidated by the discursive practices of the middle class and the state administrators, the idea of accountability no doubt is a significant stretch. The near-impossible feat of transforming accountability from an idea to practice for the urban poor dominated much of the discussion at Janaagraha. Still it was never considered an option. It remained mainly as a rhetorical device to keep the urban poor within the purview of the project as is expected of any development project and to secure funding from international donors.

Note that this rhetoric of accountability espoused by Janaagraha directly aligns with the corporate rhetoric, like that of Infosys. However, both accountability and transparency have gained wide currency even in the development sphere. While the development narrative has its own power, it is also important to remember that Bangalore as a space defined by the success of IT carries an independent rhetorical force based primarily on the ideals of corporate governance. In other words, the direct thread of legitimacy that runs between Infosys and Janaagraha has momentous impact on how participation is conceived in this context.

Conclusion: BITS of Belonging

What does an ethnography of neoliberalism look like? At the end and through the years of completing this work and thinking through the issues, I would argue that while it belies a clear answer, it also evades a linear path, both methodologically and analytically. This refusal of the linear is integral to the experience and practices of neoliberalism and also the sites through which the anthropologist builds his or her argument. The sites, such as the slums, are not necessarily external to IT; they are in a way non-IT. In being the "non," the slum mirrors the IT aspiration and in mirroring becomes its integral reflection, just with the images reversed. In the reversal, I argue, lies the politics of externalization. Some of the central questions I delved into were the following: What are the limits of corporate ethnography beyond the ethical rhetoric of "transparency" and "accountability"? In this sense, what does the characteristic limit of such an ethnographic field indicate in terms of the politics of the ethnographic method itself? At another level, my theoretical concerns have been with issues of citizenship and belonging specifically in the postcolonial context: How can we best understand how one belongs in a postcolonial nation-state? And how one knows that one in fact truly belongs? What does an ethnography that attempts to capture the contemporaneity of this kind of belonging unveil?

While the Indian contemporary as a temporality, is multiple and contested, time and again the middle class has emerged to refashion its national trajectory. For a nation whose contemporary is forever perceived to be staggering, the question of belonging is unavoidable. This question is both intellectually and ethically demanding especially when state "corruption" is the central focus of the discourse. Corruption is not new in how citizens think of the Indian state. It continually thwarts citizenship, which is under threat of subversion. If the nation-state becomes available through the discourses of corruption, it also becomes available through notions of belonging. I suggest through this work that one way to map the contemporary is to chart the social terrain of *how* one belongs to the nation-state. Beyond its economic tenor, the success of IT in global capitalism was deployed as a tool of ethical intervention and about introducing the market as the new paradigm of belonging in the nation-state.

To reiterate, this is not an ethnography of IT alone. IT is both an object and a tool of inquiry, a prism to refract an emerging social reality. Current research on IT focuses on its "influence," which primarily catalogs the spread of computer literacy and Internet connectivity especially in rural India and how that has changed lives, which were otherwise remote and isolated. While the Internet as a medium does alter the way people see and conceive the world, it offers a limited understanding of the fundamental ways in which everyday politics is being recast in contemporary India. Rather than mapping the "influence of IT," which I think is to some extent teleological, I asked what is it *in* IT, specifically, and neoliberalism in general, that persuades the imagination of the nation? How can IT, a repertoire of technological practices, persuade new social practices based upon market values? What allows Nilekani and Murthy to prescribe their model of corporate governance to disassemble and reassemble public governance and citizenship? How does their thinking recruit the wider middle class, as in Janaagraha, once again after the nationalist struggle as vanguards of change?

The corporate governance model based on *accountability* and *transparency* followed by Infosys normalizes the market as the neutral tool through which citizens can belong as consumers. India, as a developing country, is the not unique in initiating economic reforms. Most of the literature on liberalization focuses on analyzing the effect of these policies in terms of whether it augmented or stifled economic growth. I argue that to confine the analysis to a balance sheet, overlooks the actual processes through which these policies affect everyday lives of citizens, their thinking and their practices. My work in Bangalore with the IT industry and their associate NGOs, like Janaagraha, reveals the nuanced modes through which a critique of the Indian state and citizenship steadily emerged. Employing the ethnographic method, I argued that the policies of liberalization move far beyond calculable economic models and become intensely entwined in the politics and ethics of imagining a "new" India.

The seeming neutrality of the market depoliticizes belonging in the nation-state. While birthright continues be the legal basis of citizenship, as anthropologists have variously shown, it is seldom a guarantor of one's rights.

Citizenship in this sense is a continuum of increasing preferences where various groups contest to assert their rights. In a democracy, the assertion and exertion of rights is the fundamental expression of politics. With the advent of the market, citizenship rights vis-à-vis the state are not only deemed peripheral but also obsolete in the current phase of globalization. As I have discussed, employees in the IT industry take an active interest in the U.S. presidential elections, whereas seldom vote in or attend to local elections. In most gated communities, where IT professionals reside, the state is variously bypassed and subverted, water supply being the paradigmatic instance. In these islands of privately constructed "happiness," the retreat from the state and the wider society is unmistakable.

The willful retreat desensitizes the middle class to the social inequality that is surfacing in newer dimensions in the post-liberalization phase. This is a significant departure from the middle-class leadership during the nationalist struggle, which was by far more attentive to social issues. On the other hand, the middle class, as I have argued, is not completely oblivious to the underprivileged sections of the society. Rather, there is interest in inducting the urban poor within the market paradigm but through a pedagogic mission of transforming the "masses" into rational market operators. In the rush to establish the market, the state social welfare schemes, which till now have been crucial safety net for the poor, are thus considered redundant and inhibitors to global success. In other words, the apparent attempt to incorporate the urban poor at the end is a clear politics of exclusion.

The computing logic, which is analogous to the Infosys-Janaagraha initiative, renders social inequality invisible. This invisibility was apparent in the case of the urban poor, where several questions were raised regarding the viability of the market for the economically underprivileged groups and the presence of international donor agencies to decree the state's transition to a market paradigm. Yet these questions were overlooked time and again by Janaagraha as remains of the past vis-à-vis the promise of the market. As I have argued, the IT industry sees the past as an enduring burden that needs to be shed. The software code also archives past information, but the thrust of every program is to temporally move a transaction or a search forward.

An interesting discovery in this ethnography was the incongruous configuration between the IT-Janaagraha neoliberal narrative and anthropology regarding the West. Johannes Fabian asks, "What is that anthropologists try to catch with their manifold and muddled use of time?" In what he calls the "problem of coevalness," Fabian argues, "If we remember, the history of our discipline, it is in the end about the relationship between the West and the Rest" (1983: 28). A similar anxiety about coevalness between India and the West is apparent in the Infosys-Janaagraha leadership. The market paradigm in this sense is not just a panacea for a developing country. It is also a tool to negate the denial of coevalness that makes the weight of history unbearable.

Yet, alongside this negation, it is also necessary to preserve a niche place for India as the software destination in global capitalism. Therefore, it is not surprising that NASSCOM at one point equated India with IT in a tab titled "India Is IT" on its website. Thus, two issues confront us: The first is India's reliance on the West for sustaining the IT industry; conversely, IT seeks to normalize the West through this sustenance.

As I was completing this work, another citizen's organization has come together in Bangalore, called the Bangalore Political Action Committee (B.PAC).[1] Alongside sports and entertainment celebrities, it comprises several entrepreneurs such as Kiran Majumdar Shaw, CEO of Biocon, and Mohandas Pai, chairperson of the board of Manipal Global Education Services Private Limited, who was with Infosys earlier. The "PAC" segment of the title unmistakably conveys that the initiative is meant to influence politics through corporate funds. B.PAC had endorsed fourteen candidates in the last state elections, of them five won.[2] The neoliberal stirrings that started in the early 1990s to recast the notion of citizenship and belonging has developed entrenched roots in the city. One may not see the continuity of the BATF over the years, but its reincarnations are strong and its memory is enduring. In this vein I want to leave the reader with an excerpt from a recent article written by a lake activist titled "Bengaluru: Water Matters" that was published on the B.PAC website:

A household in a slum pays Rs 2–5 for a pot of water; a resident in an apartment block buys tanker supplied water (6000 litres) at Rs 300–800 and a household with the BWSSB connection pay Rs 71 KL of metered water. The glaring disparity is not just unfair but undemocratic too, why should households of the same city have to pay different cost for water they consume?[3]

1. Available at http://www.bpac.in/ (accessed May 15, 2014).
2. Available at http://www.thehindu.com/news/cities/bangalore/bpac-endorses-14-candidates/article4661314.ece (accessed May 15, 2014).
3. "Bengaluru: Water Matters," available at http://www.bpac.in/bengaluru-water-matters/ (accessed April 24, 2015).

Bibliography

Anand, N. 2011. "Pressure: The Polytechnics of Water Supply in Mumbai." *Cultural Anthropology* 26 (4): 542–564.

Anderson, B. 1991. *Imagined Communities: Reflections on the Origin and Spread of Nationalism*. London: Verso.

Aneesh, A. 2006. *Virtual Migration: The Programming of Globalization*. Durham, NC: Duke University Press.

Appadurai, A. 1996. *Modernity at Large: Cultural Dimensions of Globalization*. Minneapolis: University of Minneapolis Press.

Arora, A., and A. Gambardella. 2005. *From Underdogs to Tigers: The Rise and Growth of the Software Industry in Brazil, China, India, Ireland and Israel*. Oxford: Oxford University Press.

Bakker, K. 2005. "Neoliberalizing Nature? Market Environmentalism in Water Supply in England and Wales." *Annals of the Association of American Geographers* 95 (3): 542–565.

———. 2010. *Privatizing Water: Governance, Failure and the World's Urban Water Crisis*. Ithaca, NY: Cornell University Press.

Baviskar, A., ed. 2007. *Waterscapes: The Cultural Politics of a Natural Resource*. New Delhi: Orient Longman.

Baviskar, B. S. 2001. "NGOs and Civil Society in India." *Sociological Bulletin* 50 (1): 3–15.

Beck, U. (1986) 1992. *Risk Society: Towards a New Modernity*. Thousand Oaks, CA: Sage.

Berthoud, G. 1991. "Market." In *The Development Dictionary: A Guide to Knowledge as Power*, edited by W. Sachs, 70–87. London: Zed Books.

Bhabha, H. 1994. *The Location of Culture*. London: Routledge.

Biao, X. 2007. *Global Body Shopping: An Indian Labor System in the Information Technology Industry*. Princeton, NJ: Princeton University Press.

Biehl, J. 2005. *Vita: Life in a Zone of Social Abandonment.* Berkeley: University of California Press.

Boym, S. 2001. *The Future of Nostalgia.* New York: Basic Books.

Castells, M. 1989. *The Informational City.* London: Blackwell.

Chakrabarty, D. 1992. "Postcoloniality and the Artifice of History: Who Speaks for 'Indian' Pasts?" *Representations* 37:1–26.

——. 2000. *Provincializing Europe: Postcolonial Thought and Historical Difference.* Princeton, NJ: Princeton University Press.

——. 2002. *Habitations of Modernity: Essays in the Wake of Subaltern Studies.* Chicago: University of Chicago Press.

Chatterjee, P. 1993a. *Nationalist Thought and the Colonial World: A Derivative Discourse.* 2nd ed. Minneapolis: University of Minnesota Press.

——. 1993b. *The Nation and Its Fragments: Colonial and Postcolonial Histories.* Princeton, NJ: Princeton University Press.

——. 2002. *A Princely Impostor? The Strange and Universal History of the Kumar of Bhawal.* Princeton, NJ: Princeton University Press.

——. 2006. *Politics of the Governed: Reflections on Popular Politics in Most of the World.* Leonard Hastings Schoff Lectures. New York: Columbia University Press.

Chhibber, P., H. Shah, and R. Verma. 2014. "Does Corruption Influence Voter Choice?" *The Hindu*, May 27. Available at http://www.thehindu.com/todays-paper/tp-opinion/does-corruption-influence-voter-choice/article6051276.ece.

Chidambaram, S. 2014. "The Play in the States: The Indian Voters' 2014 Mandate." *Economic and Political Weekly* 49 (30): 22–24.

Cohen, L. 1999. "Where It Hurts: Indian Material for an Ethic of Organ Transplantation." *Daedalus* 128 (4): 135–164.

Coles, A., and T. Wallace, eds. 2005. *Gender, Water and Development.* Oxford: Berg.

Comaroff, J. 1992. *Ethnography and the Historical Imagination.* Boulder, CO: Westview.

Cooper, F., and A. Stoler, eds. 1997. *Tensions of Empire: Colonial Cultures in a Bourgeois World.* Berkeley: University of California Press.

Curtis, B., B. Hefley, and S. Miller. 2009. *People Capability Maturity Model (P-CMM) Version 2.0.* 2nd ed. Pittsburgh, PA: Carnegie Mellon University, Software Engineering Institute.

Daniel, V. 1996. *Charred Lullabies: Chapters in an Anthropography of Violence.* Princeton, NJ: Princeton University Press.

Das, V. 1999. "Public Good, Ethics, and Everyday Life: Beyond the Boundaries of Bioethics." *Daedalus* 128 (4): 99–133.

Das, V., and D. Poole. 2004. *Anthropology at the Margins of the State.* School of American Research Advanced Seminar Series. Santa Fe, NM: SAR Press.

Dasgupta, S. 2010. "Transactions in Transparency: Water and the Market Paradigm in the Indian Silicon Plateau." *Anthropology News* 51 (1): 16.

——. 2012. "Rethinking Participation: Water, Development and Democracy in Bangalore." *Journal of South Asian Studies* 35 (3): 520–545.

Deshpande, R. S. 2002. "Suicide by Farmers in Karnataka: Agrarian Distress and Possible Alleviatory Steps." *Economic and Political Weekly* 37 (26): 2601–2610.

Devi, S. U. 2002. "Information Technology and Asian Women in US." *Economic and Political Weekly* 37 (43): 4421–4428.

Dirks, N. 2001. *Castes of Mind: Colonialism and the Making of Modern India.* Princeton, NJ: Princeton University Press.

Dreze, J. 2014. "The Gujarat Muddle." *The Hindu.* Available at http://www.thehindu .com/opinion/op-ed/the-gujarat-muddle/article5896998.ece.

Dumit, J. 2000. *Living and Working with the New Medical Technologies: Intersections of Inquiry.* Cambridge: Cambridge University Press.

Dumont, L. 1970. *Homo Hierarchicus: The Caste System and Its Implications.* Chicago: University of Chicago Press.

Elyachar, J. 2012. "Before (and after) Neoliberalism: Tacit Knowledge, Secrets of the Trade, and the Public Sector in Egypt." *Cultural Anthropology* 27 (1): 76–96.

Englund, H. 2006. *Prisoners of Freedom: Human Rights and the African Poor.* California Series in Public Anthropology. Berkeley: University of California Press.

Epstein, S. 1996. *Impure Science: AIDS, Activism, and the Politics of Knowledge.* Berkeley: University of California Press.

Fabian, J. 1983. *Time and the Other: How Anthropology Makes Its Object.* New York: Columbia University Press.

Fanon, F. 1994. *Black Skin, White Masks.* New York: Grove Press.

Farmer, P. 1999. *Infections and Inequalities: The Modern Plagues.* Berkeley: University of California Press.

———. 2006. *AIDS and Accusation: Haiti and the Geography of Blame.* Berkeley: University of California Press.

Feldhaus, A. 1995. *Water and Womanhood: Religious Meanings of Rivers in Maharashtra.* Oxford: Oxford University Press.

Feldman, L. 2000. *Freedom as Motion.* Lanham, MD: University Press of America.

Fernandes, L. 2006. *India's New Middle Class: Democratic Politics in the Era of Economic Reform.* Minneapolis: University of Minnesota Press.

Fischer, M. J. 1999. "Emergent Forms of Life: Anthropologies of Late or Postmodernities." *Annual Review of Anthropology* 28:455–478.

———. 2005. "Technoscientific Infrastructures and Emergent Forms of Life: A Commentary." *American Ethnologist* 107 (1): 55–61.

Fischer, M. J., and G. Marcus. 1999. *Anthropology as Cultural Critique.* Chicago: University of Chicago Press.

Fleck, L. 1979. *Genesis and Development of a Scientific Fact.* Chicago: University of Chicago Press.

Franceys, R. 2008. "GATS, Privatization and Institutional Development for Urban Water Provision: Future Postponed." *Progress in Development Studies* 8 (1): 45–58.

Friedman, T. L. 2005. *The World Is Flat: A Brief History of the Twenty-First Century.* New York: Farrar, Straus and Giroux.

Fuller, C. J., and H. Narasimhan. 2006. "Engineering Colleges, 'Exposure' and Information Technology: Professionals in Tamil Nadu." *Economic and Political Weekly* 41 (3): 258–262, 288.

———. 2007. "Information Technology Professionals and the New-Rich Middle Class in Chennai (Madras)." *Modern Asian Studies* 41 (1): 121–150.

———. 2010. "Traditional Vocations and Modern Professions among Tamil Brahmans in Colonial and Post-colonial South India." *Indian Economic and Social History Review* 47 (4): 473–496.

———. 2014. *Tamil Brahmins: The Making of a Middle Caste*. Chicago: University of Chicago Press.

Gandhi, M. K. 2007. *Hind Swaraj and Other Writings: Centenary Editions*. Edited by Anthony J. Parel. Cambridge Texts in Modern Politics. Cambridge: Cambridge University Press.

Gandy, M. 2008. "Landscapes of Disaster: Water, Modernity, and Urban Fragmentation in Mumbai." *Environment and Planning A* 40:108–130.

Ganti, T. 2014. "Neoliberalism." *Annual Review of Anthropology* 43:89–104.

Giddens, A. 1991. *Modernity and Self Identity*. Stanford, CA: Stanford University Press.

Girdner, E. J. 1987. "Economic Liberalization in India: The New Electronics Policy." *Asian Survey* 27 (11): 1188–1204.

Goldman, M. 2006. *Imperial Nature: The World Bank and Struggles for Social Justice in the Age of Globalization*. New Haven, CT: Yale University Press.

Grieco, J. 1984. *Between Dependency and Autonomy: India's Experience with the International Computer Industry*. Berkeley: University of California Press.

Guha, R., ed. 1982. *Subaltern Studies: I. Writings on South Asian History and Society*. Delhi: Oxford University Press.

———. 1997. *Dominance without Hegemony: History and Power in Colonial India*. Cambridge, MA: Harvard University Press.

Gupta, A. 2005. "Narratives of Corruption." *Ethnography* 6 (1): 5–34.

Hansen, T. B. 2001. *Wages of Violence: Naming and Identity in Postcolonial Bombay*. Princeton, NJ: Princeton University Press.

———. 2005. "Sovereign beyond the State: On Legality and Authority in Urban India." In *Sovereign Bodies: Citizens, Migrants and the State in Postcolonial Worlds*, edited by T. B. Hansen and F. Stepputat, 169–191. Princeton, NJ: Princeton University Press.

Haraway, D. 1991. *Simians, Cyborgs and Women: The Reinvention of Nature*. New York: Routledge.

———. 1997. *Modest_Witness@Second_Millennium. FemaleMan_Meets_OncoMouse: Feminism and Technoscience*. New York: Routledge.

Hardgrave, R. L., and S. A. Kochanek. 1986. *India: Government and Politics in a Developing Nation*. New York: Harcourt.

Harding, S. 1991. *Whose Science? Whose Knowledge? Thinking from Women's Lives*. Buckingham, UK: Open University Press.

Harvey, D. 1991. *The Condition of Postmodernity: An Enquiry into the Origins of Cultural Change*. London: Wiley-Blackwell.

———. 2007. *A Brief History of Neoliberalism*. Oxford: Oxford University Press.

Heeks, R. 1996. *India's Software Industry: State Policy, Liberalization and Industrial Development*. New Delhi: Sage.

Heitzman, J. 2004. *Network City: Planning the Information Society in Bangalore*. New Delhi: Oxford University Press.

Holler, D., and C. Shore, eds. 2005. *Corruption: Anthropological Perspectives*. Anthropology, Culture and Society. Ann Arbor, MI: Pluto Press.

Holston, J., ed. 1989. *Cities and Citizenship*. Durham, NC: Duke University Press.

Iyer, R. 2003. *Water: Perspectives, Issues, Concerns*. New Delhi: Sage.

Jaffrelot, C. 2007. *Hindu Nationalism: A Reader*. Princeton, NJ: Princeton University Press.

Jaglin, S. 2002. "The Right to Water versus Cost Recovery: Participation, Urban Water Supply and the Poor in Sub-Saharan Africa." *Environment and Urbanization* 14 (1): 231–245.

Jameson, F. 1991. *Postmodern or the Logic of Late Capitalism.* Durham, NC: Duke University Press.

Kelty, C. 2008. *Two Bits: The Cultural Significance of Free Software.* Durham, NC: Duke University Press.

Khandekar, A. 2013. "Education Abroad: Engineering, Privatization and the New Middle Class." *Engineering Studies* 5 (3): 179–198.

Knorr-Cetina, K. 1999. *Epistemic Cultures: How the Sciences Make Knowledge.* Cambridge, MA: Harvard University Press.

Krishna, A., and V. Brihmadesam. 2006. "What Does It Take to Become a Software Professional?" *Economic and Political Weekly* 41 (30): 3307–3314.

Kumar, N. 2001. "Indian Software Industry Development: International and National Perspective." *Economic and Political Weekly* 36 (45): 4278–4290.

Lahiri-Dutt, K., ed. 2006. *Fluid Bonds: Views on Gender and Water.* Calcutta: Stree.

Lakha, S. 1990. "Growth of Computer Software Industry in India." *Economic and Political Weekly* 25:49–56.

———. 1994. "The New International Division of Labor and the Indian Computer Software Industry." *Modern Asian Studies* 28 (2): 381–408.

Lateef, A. 1996. *Linking Up with the Global Economy: A Case Study of the Bangalore Software Industry.* Geneva: International Institute of Labor Studies.

Latour, B., and S. Woolgar. 1986. *The Construction of Scientific Facts.* Princeton, NJ: Princeton University Press.

Lema, R., and B. Hesbjerg. 2003. *The Virtual Extension: A Search for Collective Efficiency in the Software Cluster in Bangalore.* Roskilde, Denmark: Roskilde University, Public Administration and Public Economics and International Development Studies.

Liang, L. n.d. "The Other Information City." Available at http://www.t0.or.at/wio/down loads/india/liang.pdf (accessed September 22, 2006).

Lock, M. 1993. *Encounters with Aging: Mythologies of Menopause in Japan and North America.* Berkeley: University of California Press.

Lowy, I. 2000. *Living and Working with the New Medical Technologies: Intersections of Inquiry.* Cambridge: Cambridge University Press.

Lyotard, J. F. (1979) 1984. *The Postmodern Condition: A Report on Knowledge.* Minneapolis: University of Minnesota Press.

Marcus, G. E. 1995. *Technoscientific Imaginaries: Conversations, Profiles and Memoirs.* 2nd ed. Chicago: University of Chicago Press.

Marriott, M., ed. 1955. *Village India: Studies in the Little Community.* Chicago: University of Chicago Press.

Mawdsley, E. 2012. *From Recipients to Donors: Emerging Powers and the Changing Development Landscape.* London: Zed Books.

Mazzarella, W. 2005. "Indian Middle Class." In *South Asia Keywords,* edited by R. Dwyer. Available at http://www.soas.ac.uk/csasfiles/keywords/Mazzarella-mid dleclass.pdf.

McDonald, D., and G. Ruiters. 2005. *The Age of Commodity: Water Privatization in Southern Africa.* London: Earthscan.

McFarlene, C. 2008. "Sanitation in Mumbai's Informal Settlements: State, 'Slum' and Infrastructure." *Environment and Planning A* 40 (1): 88–107.

Mehta, U. S. 1999. *Liberalism and Empire: A Study in Nineteenth-Century British Liberal Thought.* Chicago: University of Chicago Press.

Mitchell, T. 2006. "Society, Economy and the State Effect." In *The Anthropology of the State: A Reader,* edited by A. Sharma and A. Gupta, 169–186. Oxford: Blackwell.

Moore, S. 2000. "Offshore Outsourcing: India as an Oasis during Application Development Drought." *Giga Information Group's Research Digest* 3 (8).

Murthy, N. 2009. *A Better India: A Better World.* London: Penguin Global.

Nageshwar, K. 2014. "Chandrababu Naidu's Comeback." *Economic and Political Weekly* 49 (28): 31–33.

Nair, J. 2000. "Singapore Is Not Bangalore's Destiny." *Economic and Political Weekly* 35 (18): 1512–1514.

———. 2005. *The Promise of the Metropolis: Bangalore's Twentieth Century.* New Delhi: Oxford University Press.

Nanda, M. 2003. *Prophets Facing Backward: Postmodern Critiques of Science and Hindu Nationalism in India.* New Brunswick, NJ: Rutgers University Press.

Nandy, A. 2002. *Time Warps: Silent and Evasive Pasts in Indian Politics and Religion.* New Brunswick, NJ: Rutgers University Press.

NASSCOM. 2002. *The IT Industry in India—Strategic Review.* New Delhi: National Association of Software and Service Companies.

NASSCOM-McKinsey. 2002. *NASSCOM-McKinsey Report 2002: Strategies to Achieve the Indian IT Industry's Aspiration.* New Delhi: National Association of Software and Service Companies.

Nayar, B. R., and T. V. Paul. 2003. *India in the World Order: Searching for Major Power Status.* Cambridge: Cambridge University Press.

Nietzsche, F. 1989. *Genealogy of Morals.* Translated and edited by Walter Kauffman. New York: Vintage.

Nilekani, N. 2009. *Imagining India: The Idea of a Renewed Nation.* New York: Penguin.

"Onwards to the Neo-economy." 2014. *Economic and Political Weekly* 49 (28): 7–8.

Ortner, S. 2011. "On Neoliberalism." *Anthropology of This Century.* Available at http://aotcpress.com/articles/neoliberalism/.

Oza, R. 2006. *The Making of Neoliberal India: Nationalism, Gender, and the Paradoxes of Globalization.* New York: Routledge.

Pani, N. 2006. "Icons and Reform Politics in India: The Case of S. M. Krishna." *Asian Survey* 46 (2): 238–256.

Parekh, B. 1999. *Colonialism, Tradition, and Reform: An Analysis of Gandhi's Political Discourse.* New Delhi: Sage.

Parthasarathy, B. 2004. "India's Silicon Valley or Silicon Valley's India? Socially Embedding the Computer Software Industry in Bangalore." *International Journal of Urban and Regional Research* 28 (3): 664–685.

Patel, R. 2010. *Working the Night Shift: Women in India's Call Center Industry.* Stanford, CA: Stanford University Press.

Patnaik, N. 2014. "India's Shift to the Right." *Economic and Political Weekly* 49 (29): 22–24.

Petryna, A. 2002. *Life Exposed: Biological Citizens after Chernobyl.* Princeton, NJ: Princeton University Press.

Poster, W. 2013. "Hidden Sides of the Debt Economy: Emotions, Outsourcing and Indian Call Centers." *International Journal of Comparative Sociology* 54 (3): 205–227.

Prakash, G., ed. 1995. *After Colonialism: Imperial Histories and Postcolonial Displacements*. Princeton, NJ: Princeton University Press.

———. 1999. *Another Reason: Science and the Imagination of Modern India*. Princeton, NJ: Princeton University Press.

Pressman, R. 2009. *Software Engineering: A Practitioners' Approach*. 7th ed. New York: McGraw-Hill Science/Engineering/Math.

Rabinow, P., ed. 1984. *Foucault Reader*. New York: Pantheon.

———. 1999. *French DNA: Trouble in Purgatory*. Chicago: University of Chicago Press.

Radhakrishnan, S. 2011. *Appropriately Indian: Gender and Culture in a New Transnational Class*. Durham, NC: Duke University Press.

Raffles, H. 2007. "Fluvial Intimacies." In *Waterscapes: The Cultural Politics of a Natural Resource*, edited by A. Baviskar, 314–339. New Delhi: Permanent Black.

Ranganathan, M., L. Kamath, and V. Baindur. 2009. "Piped Water Supply to Greater Bangalore: Putting the Cart before the Horse?" *Economic and Political Weekly* 44 (33): 53–62.

Rapp, R. 1999. *Testing Women, Testing the Fetus: The Social Impact of Amniocentesis in America*. New York: Routledge.

Rath, S., and M. Rao. 2005. *Study of Municipal Finances in Karnataka, 2001–2005*. Bangalore: Centre for Budget and Policy Studies. Available at http://www.cbps .in/wp-content/themes/cbps/pdf/Study%20of%20Municipal%20Finances%20 in%20Karnataka%202001-2005.pdf.

Redfield, R. 1956. *Peasant, Society and Culture*. Chicago: University of Chicago Press.

Reverby, S. 2000. *Tuskegee's Truths: Rethinking the Tuskegee Syphilis Study*. Chapel Hill: University of North Carolina Press.

Rheinberger, H.-J. 1995. *Concepts, Theories and Rationality in the Biological Sciences*. Pittsburgh: University of Pittsburgh Press.

Rose, N. 1999. *Powers of Freedom: Reframing Political Thought*. Cambridge: Cambridge University Press.

Roy, A., and A. Ong, eds. 2012. *Worlding Cities: Asian Experiments and the Art of Being Global*. Chichester, UK: Wiley-Blackwell.

Roy, S. 2014. "Being the Change: The Aam Admi Party and the Politics of the Extraordinary in Indian Democracy." *Economic and Political Weekly* 49 (15): 45–54.

Sachs, J., A. Varshney, and N. Bajpai, eds. 1999. *India in the Era of Economic Reforms*. New Delhi: Oxford University Press.

Said, E. 1993. *Culture and Imperialism*. New York: Knopf.

Saikia, A. 2014. "A Look at the Economics behind BJP's Victory." *Economic and Political Weekly* 49 (29): 24–26.

Sangameswaran, P., R. Madhav, and C. D'Rozarion. 2008. "24/7 'Privatisation' and Water Reform: Insights from Hubli-Dharwad." Special issue, *Economic and Political Weekly* 43 (14): 60–67.

Saran, A. K. 1962. "Review of Contributions to Indian Sociology no. IV." *Eastern Anthropology* 15 (1): 53–68.

Sassen, S. 1991. *The Global City: New York, London, Tokyo*. Princeton, NJ: Princeton University Press.

———. 1998. *Globalization and Its Discontents: Essays on the New Mobility of People and Money*. New York: New Press.

———. 2005. "Global City: Introducing a Concept." *Brown Journal of World Affairs* 11 (2): 27–43.

Saxenian, A. 2000. "Bangalore: The Silicon Valley of Asia?" Paper presented at the Conference on Indian Economic Prospects: Advancing Policy Reform, Center for Research on Economic Development and Policy Reform, Stanford University, Stanford, CA.

Scheper-Hughes, N. 1993. *Death without Weeping: The Violence of Everyday Life in Brazil*. Berkeley: University of California Press.

Scrase, T. 2006. "The New Middle Class in India: A Reassessment." Paper presented at the Sixteenth Biennial Conference of the Asian Studies Association of Australia, Wollongong, June 26–29.

Shapin, S. 1995. *A Social History of Truth: Civility and Science in Seventeenth-Century England*. Chicago: University of Chicago Press.

Shapin, S., and S. Schaffer. 1985. *Leviathan and the Air-Pump: Hobbes, Boyle, and the Experimental Life*. Princeton, NJ: Princeton University Press.

Sharma, R. N., and A. Bhide. 2005. "World Bank Funded Slum Sanitation Program in Mumbai: Participatory Approach and Lessons Learnt." *Economic and Political Weekly* 40 (17): 1784–1789.

Shiva, V. 2008. "Seed Monopolies, Genetic Engineering and Farmers' Suicide." Available at http://www.whale.to/b/shiva.pdf.

Sitapati, V. 2011. "What Anna Hazare's Movement and India's New Middle Classes Say about Each Other." *Economic and Political Weekly* 46 (30): 39–44.

Spivak, G. C. 2000. "Megacity." *Grey Room* 1:8–25.

Srinivas, M. N. 1952. *Religion and Society among the Coorgs of South India*. Oxford: Oxford University Press.

———. 1997. "Practicing Social Anthropology in India." *Annual Review of Anthropology* 26:1–24.

Star, S. L. 1999. "The Ethnography of Infrastructure." *American Behavioral Scientist* 43 (November–December): 377–391.

Stoler, A. 2001. "Tense and Tender Ties: The Politics of Comparison in North American History and (Post) Colonial Studies." *Journal of American History* 88 (3): 829–865.

———. 2002. *Carnal Knowledge and Imperial Power: Race and the Intimate in Colonial Rule*. Berkeley: University of California Press.

Taylor, C. 2004. *Modern Social Imaginaries*. Durham, NC: Duke University Press.

Teltumbde, A. 2014. "Saffron Neoliberalism." *Economic and Political Weekly* 49 (31): 10–11.

Thompson, E. P. 1967. "Time, Work-Discipline, and Industrial Capitalism." *Past and Present* 38 (1): 56–97.

Traweek, S. 1992. *Beamtimes and Lifetimes: The World of High Energy Physicists*. Cambridge, MA: Harvard University Press.

Upadhya, C. 2004. "A New Transnational Capitalist Class? Capital Flows, Business Networks and Entreprenuers in the Indian Software Industry." *Economic and Political Weekly* 39 (48): 5141–5143.

Upadhya, C., and A. R. Vasavi. 2008. *In an Outpost of the Global Economy: Work and Workers in India's Information Technology Industry*. New Delhi: Routledge.

USAID. 2003. *Pooled Finance Framework for Greater Bangalore Water Supply and Sewerage Project for Eight Cities in Bangalore Metropolitan Area, Project Development Report.* Bangalore: Indo-USAID Financial Institutions Reform and Expansion (FIRE) Project.

van Jaarsveld, D., and W. Poster. 2013. "Call Centers: Emotional Labor over the Phone." In *Emotional Labor in the 21st Century: Diverse Perspectives on Emotion Regulation at Work,* edited by A. Grandey, J. Diefendorff, and D. Rupp, 153–174. New York: Routledge.

Vasavi, A. R. 1999. "Agrarian Distress in Bihar: Market, States and Suicides." *Economic and Political Weekly* 34 (32): 2263–2268.

Von Schnitzler, A. 2010. "Gauging Politics: Water, Commensuration and Citizenship in Post-Apartheid South Africa." *Anthropology News* 51 (1): 7–9.

Walters, V. 2013. *Water, Democracy and Neoliberalism in India: The Power to Reform.* Routledge Contemporary South Asia Series. Oxford: Routledge.

Williams, R. 1977. *Marxism and Literature.* New York: Oxford University Press.

Young, A. 1995. *The Harmony of Illusions: Inventing Post-traumatic Stress Disorder.* Princeton, NJ: Princeton University Press.

Index

Aadhaar Scheme, 17
Aam Aadmi Party (AAP; Party of the Ordinary Man), 19–20
accountability: as corporate value, 9; as market ideal, 3; in mining industry, 43; as model for government, 67, 159, 187–188; as national ideal, 149; as neutral tool, 190; transferring to urban poor, 187; as universal human value, 159; and Water Project, 36
Adhar Trust, 49
Africa, sub-Saharan water privatization, 186–187
agriculture: commercialization of, 23; foreign capital and, 68n6; and illegal land transfers to developers, 73; India shifting from, 33; and land converted to IT parks, 50
Ahluwalia, Montek Singh, 41–42
Alandur project, Tamil Nadu, 88–89
Anand, Nikhil, 12, 15, 178
Anderson, Benedict, 130, 144, 149
Andhra Pradesh, 1, 24
Aneesh, Aneesh, 26, 141–142, 147n11
Anganwadi (courtyard shelter program), 174–175
Appadurai, Arjun, 147

Appropriately Indian (Radhakrishnan), 27
APSA (Association for Promoting Social Action), 164–167
Arora, Ashish, 112
Association for Promoting Social Action (APSA), 164–167
awards, as differentiation from competition, 114

Bach, Jonathan, 12
Bachelard, Gaston, 173
Baindur, Vinay, 70–71, 73, 110
Bakker, Karen, 67, 71n10
Bangalore: Cantonment, 64; demographic profile of, 64; as dual city of wealth and poverty, 133; emulating today's global cities, 126; as epicenter for new India, 55; high-rises as landmarks in, 64–65; as neoliberal-ready space, 128; periphery of, 67; Returned NRI, 150–151; as Silicon Valley of India, 53, 98; software professionals in, 28; urban renewal in, 23. *See also* Greater Bangalore
Bangalore Agenda Task Force. *See* BATF
Bangalore City Commission (BCC), 82
Bangalore Development Authority (BDA), 49, 81, 86

Bangalore Mahanagar Palike. *See* BMP
Bangalore Political Action Committee
  (B.PAC), 192
Bangalore Water Supply and Sewerage
  Board (BWSSB). *See* BWSSB
BATF (Bangalore Agenda Task Force):
  discontinuation of, 50–51; evoking of
  poverty as collateral, 59; focusing on city
  infrastructure, 48–49; no elected rep-
  resentatives as members of, 127; public-
  private initiative to reform governance
  of, 11, 48; reducing public transport, 19,
  50, 61
Baviskar, B. S., 13
BBMP (Bruhat Bengaluru Mahanagara
  Palike), 83
BCC (Bangalore City Commission), 82
BDA (Bangalore Development Authority),
  49, 81, 86
Beck, Ulrich, 120
belonging: in era of neoliberalism, 4, 59,
  190; new politics of, 4; in postcolonial
  nation-state, 189–190
Berthoud, Gerald, 179
Better India, A (Murthy and Nilekani), 35
Betterment Tax, 90
Bhabha, Homi, 130, 144–145
Bharat Earth Movers, 23
Bharat Electronics, 23
Bharatiya Janata Party (BJP), 7, 15–16,
  20–22, 43
Bhide, Amita, 14
Biao, Xiang, 2, 112
Biehl, João, 159
biotechnology, and farmer suicides, 24
BJP (Bharatiya Janata Party), 7, 15–16,
  20–22, 43
Black Skin, White Masks (Fanon), 46
blueprint for governance, 4, 30, 97, 190
BMP (Bangalore Mahanagar Palike):
  administering city of Bangalore, 72;
  BBMP giving jurisdiction over Greater
  Bangalore, 84; BWSSB providing piped
  water and sewerage, 74; jurisdictional
  confusion of with CMC, 80, 83; pre-
  sumed performance of relative to CMC,
  83–84
Bodo-Muslim conflict in Assam, 56
body shopping, 2, 99
Bolivia, 186
Bomanahalli CMC, 71

bore wells, 19, 73–74, 82
Bose, Subhas Chandra, 148
Boym, Svetlana, 137–138
B.PAC (Bangalore Political Action Com-
  mittee), 192
Brazil, 125–128, 185
bribery, 19, 23, 80
Brihmadesam, Vijay, 135
Bruhat Bengaluru Mahanagara Palike
  (BBMP), 83
Bush, George, 96
business investment, Indian, U.S., and
  U.K., 111
BWSSB (Bangalore Water Supply and
  Sewerage Board; Water Board): and Act
  of 1964, 85–86; creating a water market,
  173; explaining new water connections to
  poor, 174–176; having no jurisdictional
  mandate in CMC, 74, 83; individual
  and shared connections by, 74–75, 174;
  lacking price discussions, 179–180;
  Manjunathanagara and, 160–161,
  163–164; monthly bills from as proof of
  residence, 181; and moral rights of poor,
  172–173; offices of as public space, 103;
  project implementation role of for Water
  Project, 70; supplying bulk water to
  Greater Bangalore, 82n28, 75, 165; WTP
  not specifying prices charged by, 166–167
Byatarayanapura CMC, 71

Capability Maturity Model (CMM), 103
caste and IT industry, 27–28
Castells, Manuel, 25, 133
Cauvery River, 171–173
CBPS (Center for Budget and Policy
  Studies), 83–84, 125–127
CDP (Comprehensive Development Plan), 49
Center for Budget and Policy Studies
  (CBPS), 83–84, 125–127
centralized planning: disempowering local
  control, 71; failure of, 155
Chakrabarty, Dipesh, 131
Chatterjee, Partha, 154–155, 157, 159–160
Chhibber, P., 20
Chidambaram, P., 67
Chidambaram, Soundaraya, 21
citizen participation: critical to "doing
  democracy," 159; Janaagraha statements
  on, 77–78, 91–92; limited to middle class,
  14, 165; and MoU concept, 87–88; as

Simanti Dasgupta is Associate Professor of Anthropology at the University of Dayton.

www.ingramcontent.com/pod-product-compliance
Lightning Source LLC
Chambersburg PA
CBHW020349270326
41926CB00007B/364